SOLAR ELECTRICITY

Second Edition

UNESCO ENERGY ENGINEERING SERIES
ENERGY ENGINEERING LEARNING PACKAGE

Dr Boris Berkovski
Chairperson of the Editorial Board
Director, Division of Engineering and Technology
UNESCO

Organized by UNESCO, this innovative distance learning package has been established to train engineers to meet the challenges of today and tomorrow in this exciting field of energy engineering. It has been developed by an international team of distinguished academics coordinated by Dr Boris Berkovski. This modular course is aimed at those with a particular interest in renewable energy, and will appeal to advanced undergraduate and postgraduate students, as well as practising power engineers in industry.

Solar Electricity, Second Edition
Edited by Tomas Markvart

Magnetohydrodynamic Electrical Power Generation
Hugo Messerle

Energy Planning and Policy
Maxime Kleinpeter

Ocean Thermal Energy Conversion
Patrick Takahashi and Andrew Trenka

Industrial Energy Conservation
Charles M. Gottschalk

Biomass Conversion and Technology
Charles Y. Wereko-Brobby and Essel B. Hagan

Mini Hydropower
Tong Jiandong et al.

Wind Energy Technology
John F. Walker and Nicholas Jenkins

SOLAR ELECTRICITY
Second Edition

Edited by
Tomas Markvart
University of Southampton, UK

JOHN WILEY & SONS, LTD
Chichester • New York • Weinheim • Brisbane • Toronto • Singapore

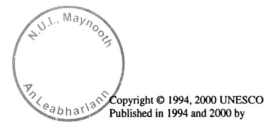

Copyright © 1994, 2000 UNESCO
Published in 1994 and 2000 by

John Wiley & Sons Ltd, The Atrium, Southern Gate,
Chichester,West Sussex PO19 8SQ, England
Telephone   (+44) 1243 779777

Email (for orders and customer service enquiries): cs-books@wiley.co.uk
Visit our Home Page on www.wileyeurope.com or www.wiley.com

Reprinted with corrections October 2000, July 2001, February 2003, March 2004,
January 2005, May 2006, February 2007, March 2008, December 2008, February 2009

Other Wiley Editorial Offices

John Wiley & Sons Inc., 111 River Street, Hoboken, NJ 07030, USA

Jossey-Bass, 989 Market Street, San Francisco, CA 94103-1741, USA

Wiley-VCH Verlag GmbH, Boschstr. 12, D-69469 Weinheim, Germany

John Wiley & Sons Australia Ltd, 33 Park Road, Milton, Queensland 4064, Australia

John Wiley & Sons (Asia) Pte Ltd, 2 Clementi Loop #02-01, Jin Xing Distripark, Singapore
129809

John Wiley & Sons Canada Ltd, 22 Worcester Road, Etobicoke, Ontario, Canada M9W 1L1

British Library Cataloguing in Publication Data

A catalogue record for this book is available from the British Library

ISBN 978-0-471-98852-6 (HB)
ISBN 978-0-471-98853-3 (PB)

Typeset in 10/12 Times by Kolam Information Services Pvt Ltd, Pondicherry, India

This book is printed on acid-free paper.

Contents

Foreword

Education applied to a whole complex of interlocking problems is the master key that can open the way to sustainable development.

A major obstacle to the development of renewable energies is the paucity of relevant information available to engineers and technicians (who too often lack the necessary know-how and skills) as well as to decision-makers and users. One of UNESCO's priorities in this domain is therefore to promote training and information aimed at sensitizing specialists and the general public to the possible uses of renewable energy sources, with particular regard to environmental concerns and the requirements of sustainable development.

The present textbook on *Solar Electricity, Second Edition* in UNESCO's Energy Engineering Series is designed not only to provide instruction for a new generation of engineers but also to foster the kind of environmental management so urgently needed throughout the world. The series as a whole is a venture involving many institutions working together – on the basis of pre-defined standards – to promote awareness of the environmental, cultural, economic and social dimensions of renewable energy issues, as well as knowledge of those technical aspects that form the core of the UNESCO Learning Package Programme.

The highly successful World Solar Summit, held in Harare, Zimbabwe, in September 1996, endorsed five activities of universal scope and value as part of the World Solar Programme 1996–2005. One of these is the *Global Renewable Energy Education and Training Programme*, of which this Learning Package constitutes a first and significant element. The UNESCO Energy Engineering Series should make a useful contribution to the World Solar Programme 1996–2005, and UNESCO is pleased to be a partner in its development.

Federico Mayor
Former Director-General, UNESCO

Preface to the First Edition

This book deals with photovoltaics: the direct conversion of light into electricity. It is based on the Short Course on Solar Energy Conversion and Applications which takes place annually at the University of Southampton but has been expanded to provide a self-contained document suitable for distance learning.

A brief overview of the subject is given in Chapter 1, devoted to the history, present status and future of this increasingly important renewable energy source.

In Chapter 2 we show how to determine the supply of solar energy for a given location. The geographical effects are analysed, and it is shown how the available solar radiation data can be used to determine the useful energy supply.

Chapter 3 explains how solar cells work and how they are manufactured. The principles of photovoltaic energy conversion are here discussed alongside the practical solar cell operation. The manufacture of the solar cells which are most commonly used at present and those entering the market—crystalline silicon and thin films—is described in some detail.

The photovoltaic power system is discussed in Chapter 4 which looks at the system structure and subsystems. This unit also includes the system sizing and design.

Applications of solar electricity are reviewed in Chapter 5, together with the economics of PV installations. Attention is focused on the various applications of photovoltaics which are relevant at present and also on those that are likely to be of increasing importance in the near future.

The environmental and social impacts of energy production are considered in Chapter 6 where the methodology for environmental assessment of energy sources is described and applied to photovoltaic power generation.

Important specialised topics are treated in Chapter 7 which includes the discussion of large photovoltaic systems, photovoltaics under concentrated sunlight, and the hydrogen economy. The preparation of this text has been a team effort. Many experts in various aspects of photovoltaics have shared their knowledge with me and, with their help, I have moulded their contributions into the format before you. All these contributions have been of a very high standard and I accept full responsibility for any inadvertent errors or ommissions in the printed text.

Professor Arthur Willoughby, Mr Fred Treble, Professor Peter Landberg and Mr John Bonda who give lectures at the Southampton course have conveyed much information to me through their lecture notes, class work and discussions, and their input is reflected in this module.

Professor Israel Dostrovsky has, through the kind gift of his book, given me much insight into the global aspects of energy production. Professor Ray Arnold has patiently explained to me the structure of electricity transmission and distribution systems—in particular, the meaning of the term grid. I am also indebted to Mr Paul Maycock, the Directorate General for Energy of the Commission of the European Communities, and all contributors of the data and photographs that are reproduced in the text.

I would like to express particular thanks to Professors Antonio Luque, Gabriel Sala, Gerardo Araujo, Eduardo Lorenzo, and all their colleagues and students, for many stimulating discussions and hospitality during my stay at the Instituto de Energia Solar in Madrid where most of this text was prepared. The help of Mr Ricardo Castrillo with tracing bibliography and of Mrs Montserrat Rodrigo with the graphics is also much appreciated.

Tomas Markvart
Madrid
June 1992

Preface to the Second Edition

The eight years since the first edition of Solar Electricity was submitted to UNESCO mark an unprecedented expansion of photovoltaics, with the world solar cell production growing threefold, and recent announcements of ambitious national and trans-national programs across several Continents. To capture the flavour of such a rapidly growing field in a textbook format has been a challenging task but, as in the first edition, this undertaking has been lightened by being able to rely on the knowledge and advice of distinguished and experienced co-authors.

The choice of new material for the second edition has been guided by the recent developments in photovoltaic science, engineering and architecture, as well as by perceived omissions in the coverage of the field in the first edition. A new section on Electrochemical Photovoltaics describes important new advances in dye-sensitised solar cells, and provides the reader with an interface to an allied subject of photochemical energy conversion. A section on PV-diesel hybrid energy systems has been added to cover a significant market sector of photovoltaic systems with application to remote area power supplies.

The second edition introduces a new method of presentation in the form of 'Boxes' which give further detail or relevant information that may be of interest but is not essential for the understanding of the main text. Thus, a box on photovoltaics in buildings describes the novel area of photovoltaic architecture which was in its infancy when the first edition went to press. The lack of suitable financing mechanisms represents arguably the most significant barrier to wider usage of solar electricity in the developing world, and a box on this topic examines the recent progress in this field. There is also a box on solar powered marine navigation aids which outlines one of the best established applications of photovoltaics in the northern latitudes.

These additions have been made possible by the efforts of an expanded group of co-authors and by the generous help of a number of persons and institutions that have made available additional materials. I am indebted to Paul Maycock of PV News for supplying detailed statistics about the photovoltaic market, to Dr Erge and colleagues of the Frauenhofer Institute for Solar Energy Systems for information about the German 1000 Roofs Programme, and to Ray Noble of BP Solarex for photographs of the G8 photovoltaic building. Akeler

Developments kindly provided information and photographs of Doxford Solar Offices. Stimulating and informative discussions with my colleagues Peter Landsberg and Bob Greef have given me much encouragement and help in the understanding of many interdisciplinary issues. My thanks and gratitude goes to my wife without whose support and patience this book would not have been completed.

But there have also been moments of sadness and sorrow as we lost two colleagues and friends whose thought and knowledge have been singular in their influence on this book. John Bonda and Bob Hill passed away in 1999 and their wisdom and expertise will be a sad loss not only to readers of Solar Electricity but to everyone in the world of photovoltaics. I hope that his Second Edition will stand as memento of their immense contributions to the science of solar energy and its applications.

Tomas Markvart
Southampton
March 2000

List of Contributors

Raymond Arnold   Siemens, Manchester, UK
Klaus Bogus   ESTEC-ESA, Noordwijk, The Netherlands
Luis Castañer   Universidad Politécnica de Cataluña, Barcelona, Spain
Andres Cuevas   Australian National University, Canberra
Philip Davies and **Juan Carlos Miñano**   Universidad Politecnica de Madrid, Spain
Miguel Angel Egido   Universidad Politecnica de Madrid, Spain
Jennyi Gregory   IT Power, Eversley, UK
Robert Hill   University of Northumbria, UK
Ben Hill   BP Solarex, Madrid, Spain
Rogelio Leal   Universidad Politécnica de Cataluña, Barcelona, Spain
Eduardo Lorenzo   Universidad Politécnica de Madrid, Spain
Sean McCarthy and **Martin Hill**   Hyperion, Cork, Ireland
Augustin McEvoy   Ecole Polytechnique Fédérale de Lausanne, Switzerland
Chem Nayar, Benjamin Wichert and **William Lawrance**   Curtin University, Perth, Australia
Oliver Paish, Bernard McNelis and **Anthony Derrick**   IT Power, Eversley, UK
Neil Ross   University of Southampton, UK
Aleksei Sorokin   Rome, Italy
Michael Specht   Center for Solar Energy and Hydrogen Research, Stuttgart, Germany
Malcolm Wannell   Trinity House Lighthouse Service, East Cowes, UK

Note on Energy Units

We have endeavoured to use the SI system of units as much as possible. However, in two instances it has been expedient (and certainly preferable from the didactic viewpoint) to employ other energy units, as defined by the SI convention.

Firstly, the usual unit of energy in solid-state physics is the electronvolt (eV). This unit is also in common use when discussing semiconductor solar cells.

Secondly, the unit watthour (Wh) and its derivatives (k Wh, etc), which are in common use in power engineering, have been used for solar irradiation. This usage is also recommended within the European Community's Solar Energy RD programme.

The following conversion factors may be found useful.

Factor	Prefix	Symbol
10^{-6}	micro	μ
10^{-3}	milli	m
10^{-2}	centi	c
10^{3}	kilo	k
10^{6}	mega	M
10^{9}	giga	G
10^{12}	tera	T
10^{18}	exa	E

$$1 \text{ W s} = 1 \text{ J}$$
$$1 \text{ Wh} = 3.6 \text{ kJ}$$
$$1 \text{ EJ} = 278 \text{ TWh}$$
$$1 \text{ eV} = 1.602 \times 10^{-19} \text{ J}$$

1

Electricity from the Sun

1.1 WHY WE NEED PHOTOVOLTAICS

The Sun has been worshipped as a life-giver to our planet since ancient times. The industrial ages gave us the understanding of sunlight as an energy source. This discovery has never been more important than now as we realise that the exploitation of fossil energy sources may be affecting the planet's ambient.

The energy supply from the Sun is truly enormous: on average, the Earth's surface receives about 1.2×10^{17} W of solar power. This means that in less than one hour enough energy is supplied to the Earth to satisfy the entire energy demand of the human population over the whole year. Indeed, it is the energy of sunlight assimilated by biological organisms over millions of years that has made possible the industrial growth as we know it today. Most of the other renewable means of power generation also depend on the Sun as the primary source: hydroelectric, wind and wave power all have the same origin (Fig. 1.1).

Figure 1.1 does not include another phenomenon whose full significance has only come to light in recent years. Some of the infrared radiation emitted by the Earth is absorbed by gases in the atmosphere and re-emitted back to the surface, maintaining the Earth's average temperature near 15 °C (Fig. 1.2). Although the precise consequences of this greenhouse effect are still subject to discussion, the global impact of carbon dioxide emissions from fossil fuel combustion is now accepted as beyond reasonable doubt. Energy sources such as photovoltaics are needed to help reduce the levels of greenhouse gases in the atmosphere and alleviate this global warming.

The history of photovoltaics takes us back over 150 years when, in 1839, Alexandre-Edmund Becquerel observed that 'electrical currents arose from certain light-induced chemical reactions' (Table 1.1). A similar effect was observed in a solid (selenium) several decades later. A comprehensive understanding of these phenomena, however, had to await the progress of science towards the quantum theory in the early parts of the nineteenth century. The development of the first solid-state devices in the late 1940s then paved the way

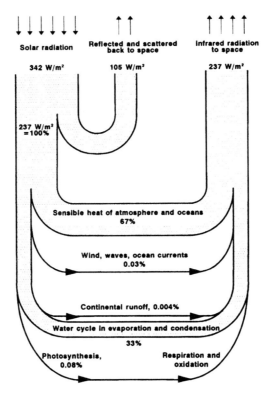

Fig. 1.1 Energy balance of the Earth. Note that the average incident solar radiation is equal to a quarter of the solar constant, to be discussed in Chapter 2 (adapted from Dostrovsky, 1988)

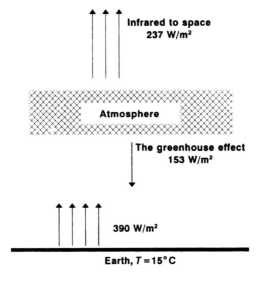

Fig. 1.2 The infrared radiation flux in the atmosphere and the greenhouse effect (adapted from J. Gribbin, Hothouse Earth, Black Swan, London, 1990)

Table 1.1 Some dates of relevance to photovoltaic solar energy conversion†

1839	Becquerel discovers the photogalvanic effect
1876	Adams and Day observe photovoltaic effect in selenium
1900	Planck postulates the quantum nature of light
1930	Quantum theory of solids proposed by Wilson
1940	Mott and Schottky develop the theory of solid-state rectifier (diode)
1949	Bardeen, Brattain and Shockley invent the transistor
1954	Chapin, Fuller and Pearson announce 6% efficient silicon solar cell
1954	Reynolds *et al.* report solar cell based on cadmium sulphide
1958	First use of solar cells on an orbiting satellite Vanguard 1

† After P. T. Landsberg and F. C. Treble, lecture notes at the Southampton Short Course on Solar Energy Conversion and Applications.

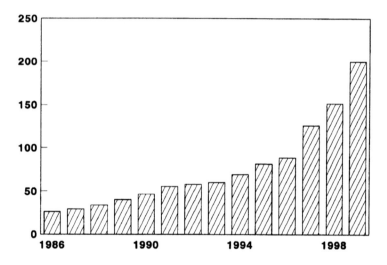

Fig. 1.3 World PV module sales (MW$_p$). (Source: P. D. Maycock, PV News.)

to the announcement of a silicon solar cell with 6% efficiency (Chapin *et al.*, 1954)—the first usable solar cell.

Solar cells did not have to wait long to find application. The year 1958 witnessed the launch of Vanguard 1, the first satellite to use electricity from the sun. The technology has been developing ever since. Much interest in solar electricity appeared particularly in the wake of the oil crisis in the early 1970s. Today, the direct conversion of light into electricity, or *photovoltaics*, is becoming accepted as an important form of power generation.

Some 200 W$_4$ of photovoltaic power modules were produced in 1999 (Fig. 1.3). The production rate has been increasing at almost 20% annually over the last decade, and is likely to reach the scale of gigawatts in the early decades of this millennium.

Photovoltaic power generation is reliable, involves no moving parts, and the operation and maintenance costs are very low. The operation of a photovoltaic system is silent, and creates no atmospheric pollution. Photovoltaic systems are modular, and can be quickly installed. Power can be generated where it is required without the need for transmission lines.

There are already a number of terrestrial applications today where photovoltaics provides a viable means of power generation. Many of these installations operate in locations where other means of electricity supply would be out of the question, providing important social benefits to rural communities throughout the world.

The cost of solar electricity already compares favourably with other small power sources such as diesel generators. The one major drawback is a high capital cost, and new financing mechanisms are being put in place to spread the payments over the system lifetime. Other innovative solutions such as photovoltaic modules integrated in the fabric of buildings reduce the marginal cost of photovoltaic energy to a minimum. The economic comparison with conventional energy sources is certain to receive a further boost as the environmental and social costs of power generation are included fully in the picture.

SUMMARY OF THE CHAPTER

The history, present status and future of photovoltaics have been outlined.

BIBLIOGRAPHY AND REFERENCES

CHAPIN, D. M., FULLER, C. S. and PEARSON, G. L., A new p-n junction photocell for converting solar radiation into electrical power, *J. Appl. Phys.* **25**, 1954: 676–677.
DOSTROVSKY, I., *Energy and the Missing Resource*, Cambridge University Press, Cambridge, 1988.
GRIBBIN, J., *Hothouse Earth*, Black Swan, London, 1990.

2

Solar Radiation

AIM

The aim of this unit is to determine the solar energy available to photovoltaic systems.

OBJECTIVES

After completing this unit you should be able to:

1. understand the nature of solar radiation at different locations on the Earth,
2. describe the apparent daily and yearly motion of the Sun,
3. show how the amount of solar radiation falling on photovoltaic panels can be determined from the measured data.

NOTATION AND UNITS

Symbol		SI unit	Other unit
B	Daily beam irradiation on a horizontal plane (monthly mean)	J/m^2	Wh/m^2
$B(\beta)$	Daily beam irradiation on a plane at angle β (monthly mean)	J/m^2	Wh/m^2
B_o	Daily extraterrestrial irradiation on horizontal plane	J/m^2	Wh/m^2
D	Daily diffuse irradiation on a horizontal plane (monthly mean)	J/m^2	Wh/m^2
$D(\beta)$	Daily diffuse irradiation on a plane at angle β (monthly mean)	J/m^2	Wh/m^2
G	Daily global irradiation on a horizontal plane (monthly mean)	J/m^2	Wh/m^2
$G(\beta)$	Daily global irradiation on a plane at angle β (monthly mean)	J/m^2	Wh/m^2

Edited from manuscripts of E. Lorenzo (Solar radiation) and L. Castañer (Photovoltaic engineering).

(Contd)

R	Radius of the Earth	m	
S	Solar constant	W/m^2	
K_T	Clearness index		
$R(\beta)$	Daily irradiation reflected from the ground onto a plane at angle β	J/m^2	Wh/m^2
α	Solar elevation		
β	Panel inclination to the horizontal plane		
δ	Solar declination		
θ_z	Zenith angle		
λ	Wavelength	m	
ϕ	Geographical latitude		
ρ	Reflectivity of the ground		
ψ	Azimuth		
ω	Hour angle		
ω_s	Sunrise hour angle		
ω_s'	Sunrise hour angle above an inclined plane		
ω_o	Minimum of ω_s and ω_s'		
	Irradiance	W/m^2	
	Spectral irradiance	$W/m^2 \mu m$ *or* $W/m^2 eV$	

Unit conversion factors
To convert a quantity given in Wh/m^2 to J/m^2, it should be multiplied by 3600.

Values of physical constants
$R = 6380\,km$
$S = 1367\,W/m^2$

2.1 INTRODUCTION

The design of a photovoltaic system relies on a careful assessment of solar radiation at a particular site. Although solar radiation data have been recorded for many locations in the world, they have to be analysed and processed before a sufficiently accurate estimate of the available solar radiation for a photovoltaic system can be made.

This chapter reviews the properties of solar radiation on Earth and outlines the calculation which must be carried out to determine the amount of radiation falling on the PV modules.

First, different components of the radiation will be introduced, and a broad assessment of the global availability of solar radiation will be made.

The astronomical relationship between the Sun and the Earth will then be described. We shall define the relevant quantities which characterise the apparent motion of the Sun.

Finally, we take a closer look at a procedure which makes it possible to determine the amount of radiation available to PV modules at an inclined position from the available solar radiation data on a horizontal surface.

2.2 ENERGY FROM THE SUN

2.2.1 Introduction

To a good approximation, the Sun acts as a perfect emitter of radiation (*black body*) at a temperature close to 5800 K. The resulting (average) energy flux incident on a unit area perpendicular to the beam outside the Earth's atmosphere is known as the *solar constant*:

$$S = 1367 \, \text{W/m}^2 \tag{2.1}$$

In general, the total power from a radiant source falling on a unit area is called *irradiance*.

The total energy flux incident on the Earth is obtained by multiplying S by πR^2, where R is the radius of the Earth. This is the area of the disk presented to the Sun's radiation by the Earth. The average flux incident on a unit surface area is then obtained by dividing this number by the total surface area of the Earth ($4\pi R^2$), giving

$$S/4 = 342 \, \text{W/m}^2 \tag{2.1a}$$

This number was used in Chapter 1 to discuss the energy balance of the Earth.

When the solar radiation enters the Earth's atmosphere (Fig. 2.1), a part of the incident energy is removed by *scattering* or *absorption* by air molecules, clouds and particulate matter usually referred to as aerosols. The radiation that

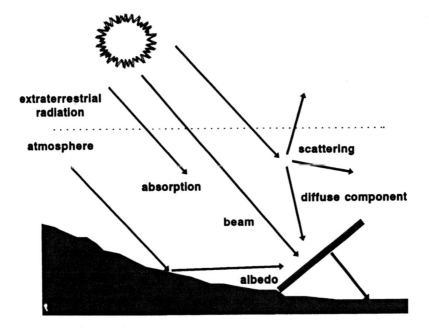

Fig. 2.1 Solar radiation in the atmosphere

is not reflected or scattered and reaches the surface directly in line from the solar disc is called *direct* or *beam radiation*. The scattered radiation which reaches the ground is called *diffuse radiation*. Some of the radiation may reach a receiver after reflection from the ground, and is called the *albedo*. The total radiation consisting of these three components is called *global*.

The amount of radiation that reaches the ground is, of course, extremely variable. In addition to the regular daily and yearly variations due to the apparent motion of the Sun, irregular variations are caused by the climatic conditions (cloud cover), as well as by the general composition of the atmosphere. For this reason, the design of a photovoltaic system relies on the input of measured data close to the site of the installation.

A concept which characterises the effect of a clear atmosphere on sunlight is the *air mass* (Fig. 2.2), equal to the relative length of the direct beam path through the atmosphere. On a clear summer day at sea level, the radiation from the sun at zenith corresponds to air mass 1 (abbreviated to AM1); at other times, the air mass is approximately equal to $1/\cos\theta_z$, where θ_z is the zenith angle which will be discussed more fully in the next section.

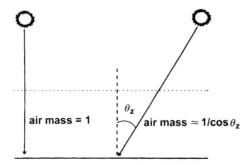

Fig. 2.2

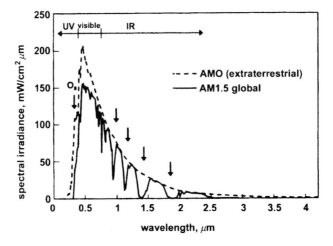

Fig. 2.3 The solar spectrum

The effect of the atmosphere (as expressed by the air mass) on the solar spectrum is shown in Fig. 2.3. The extraterrestrial spectrum, denoted by AM0, is important for satellite applications of solar cells. AM1.5 is a typical solar spectrum on the Earth's surface on a clear day which, with total irradiance of $1 \, \text{kW/m}^2$, is used for the calibration of solar cells and modules. Also shown in Fig. 2.3 are the principal absorption bands of the air molecules.

Although the global irradiance can be as high as $1 \, \text{kW/m}^2$, the available irradiance is usually considerably less than this maximum value because of the rotation of the Earth and adverse weather conditions. For illustration, Fig. 2.4 shows the mean annual irradiance in different parts of the world. The mean irradiance is highest near the latitudes of the tropics of Cancer and Capricorn, but is lower in equatorial regions on account of the cloud cover. In higher latitudes, of course, solar radiation is weaker because of low solar elevation.

Solar irradiance integrated over a period of time is called *solar irradiation*. Of particular significance in the design of photovoltaic systems is the irradiation over one day. The yearly variation of both the global and diffuse daily irradiation on a horizontal plane for four locations ranging from the Sahara desert to northern Europe is shown in Fig. 2.5. Note that the seasonal variation becomes more pronounced with increasing latitude.

The average daily solar radiation on the ground G_{av} (averaged both over the location and time of the year) can be obtained by a simple argument as follows. According to equation (2.1a) the average irradiance outside the atmosphere is equal to $342 \, \text{W/m}^2$. As we have seen in Chapter 1 and discussed above, solar

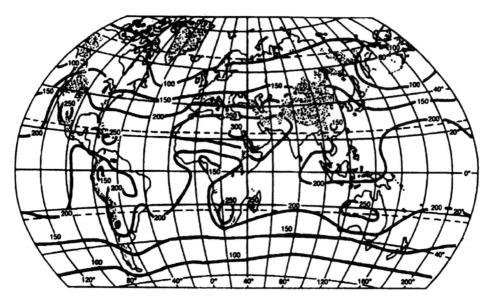

Fig. 2.4 The mean annual irradiance on a horizontal plane (in W/m^2) across the world (from F. Kreith and J. F. Kreider *Principles of Solar Engineering.*, Hemisphere Publishing Corp., New York, 1978, page 17. Reproduced with permission)

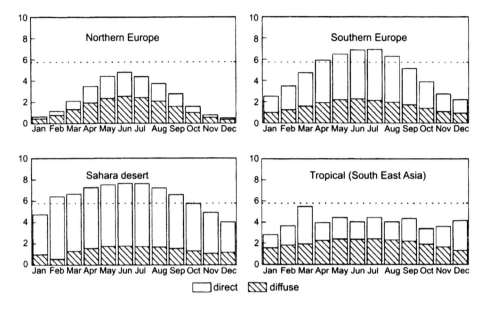

Fig. 2.5 The yearly profile of mean daily solar radiation (in kWh/m^2) as a function of geographic location. Dashed line indicates the world average

radiation observed on the Earth surface is 30% lower on account of scattering and reflection of radiation. Thus,

$$G_{av} = 0.7 \times 342 \times 24\,h = 5.75\,kWh/day$$

This value should be compared with the daily solar radiation observed in different parts of the world (Fig. 2.5).

2.2.2 Some astronomy

The Earth revolves around the Sun in an elliptical orbit (very close to a circle) with the Sun in one of the foci. The plane of this orbit is called the *ecliptic*. The time taken for the Earth to complete this orbit defines a year. The relative position of the Sun and Earth is conveniently represented by means of the *celestial sphere* around the Earth (Fig. 2.6). The equatorial plane intersects the celestial sphere in the *celestial equator*, and the polar axis in the *celestial poles*. The motion of the Earth round the Sun is then pictured by the apparent motion of the Sun in the ecliptic which is tilted at 23.45° to the celestial equator. The angle between the line joining the centres of the Sun and the Earth and the equatorial plane is called the *solar declination* and denoted by δ. This angle is zero at the *vernal* (20/21 March) and *autumnal* (22/23 September) *equinoxes*. On these days, the sun rises exactly in the east and sets exactly in the west. At the *summer solstice* (21/22 June), δ = 23.45° and at the *winter solstice* (21/23 December), δ = −23.45°.

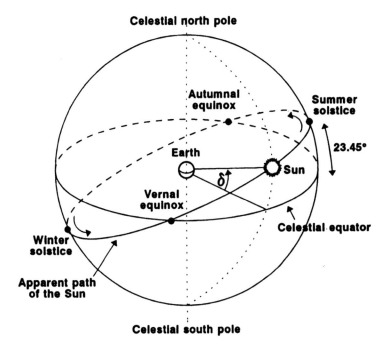

Fig. 2.6 The celestial sphere with the apparent yearly motion of the Sun

The Earth itself rotates, at the rate of one revolution per day, about the *polar axis*. The daily rotation of the Earth is depicted by the rotation of the celestial sphere about the polar axis, and the instantaneous position of the sun is described by the *hour angle* ω: the angle between the meridian passing through the sun and the meridian of the site. The hour angle is zero at solar noon and increases towards the east (Fig. 2.7(a)). For an observer on the Earth's surface at a location with geographical latitude φ, a convenient coordinate system is defined by a vertical line at the site which intersects the celestial sphere in two points, the *zenith* and the *nadir*, and subtends the angle φ with the polar axis. The great circle perpendicular to the vertical axis is the *horizon*. The angle between the Sun's direction and the horizon is the *elevation* α whose complement to 90° is the *zenith angle* θ_z. The other coordinate in this system is the *azimuth* ψ which is zero at solar noon and increases towards the east. During the daily motion, the solar declination δ can usually be assumed constant and equal to its value at midday.

The following quantities will be needed in the next section to calculate the amount of solar radiation falling on an inclined surface (Fig. 2.7(b)).

Solar declination δ, approximately given by (in radians):

$$\delta = \pi \frac{23.45}{180} \sin\left(2\pi \frac{284 + n}{365}\right) \tag{2.2}$$

(a)

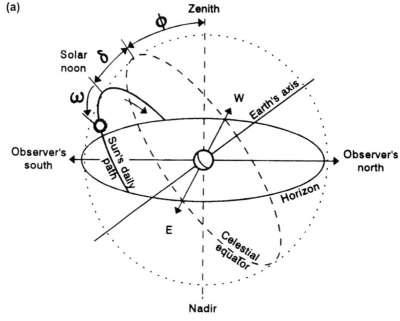

(b)

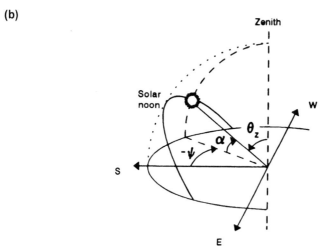

Fig. 2.7 (a) The local zenith—nadir coordinate system showing the apparent daily motion of the Sun. (b) Definition of the azimuth, solar elevation and the zenith angle

where n is the number of the day in the year ($n = 1$ on 1st January, for example).

Solar elevation α or zenith angle θ_z:

$$\sin \alpha = \sin \delta \sin \phi + \cos \delta \cos \phi \cos \omega = \cos \theta_z \qquad (2.3)$$

Solar azimuth ψ:

$$\cos \psi = (\sin \alpha \sin \phi - \sin \delta) / \cos \alpha \cos \phi \qquad (2.4)$$

These equations can be used to determine the sunrise hour angle ω_s:

$$\omega_s = \cos^{-1}(-\tan \phi \tan \delta) \qquad (2.5)$$

The sunset hour angle is then simply $-\omega_s$.

Box 2.1 Solar time

All time quantities used in this book refer to apparent solar time. Because the Sun moves along the ecliptic rather than along the equator and the distance between the Earth and the Sun is not constant, the apparent motion of the Sun contains an irregular component and the apparent solar time does not coincide with the time measured by conventional clocks. To obtain a time system where every day is of equal length, the astronomers define a fictitious body called the *mean sun* which moves with uniform speed along the equator, and whose motion is the average over the year of the true Sun's angular motion along the ecliptic. The mean solar time is then defined as the hour angle of the mean Sun. The difference between the apparent solar time and the mean time – called the *equation of time* – can be expressed mathematically as

$$E = 2.292(0.0075 + 0.1868 \cos \beta - 3.2077 \sin \beta - 1.4615 \cos 2\beta - 4.089 \sin 2\beta)$$

where the angle β is, in radians,

$$\beta = \frac{2\pi}{365} (n - 1)$$

and n is the day of the year (Fig. B2.1).

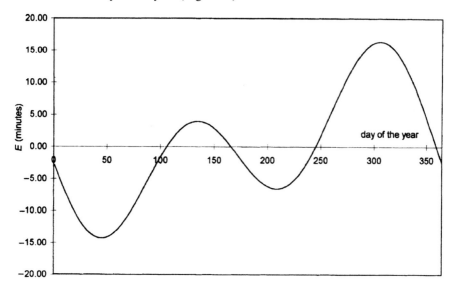

Fig. B2.1

To avoid the use of this infinity of time systems particular to each longitude, local time systems are used where the same mean time – called zone time or standard time – is used within a standard time zone. The difference between the local mean time and the standard time is easy to determine if we note that the Sun takes 4 minutes to traverse 1 degree of longitude.

The full difference between the solar time and the standard time can now be written, in minutes

$$\text{Solar time} - \text{standard time} = 4(L_{st} - L_{loc}) + E$$

where L_{st} is the standard meridian for the local time zone, and L_{loc} is the longitude of the location. (*Adapted from Duffie and Beckman, 1991*)

2.2.3 Radiation on an inclined surface

The solar radiation data, when available, are frequently given in the form of *global radiation on a horizontal surface* at the site, for example, in the form of global daily irradiation values (denoted by G) for a typical day in any one month (see, for example, Palz, 1984, and Lof *et al.*, 1966). Since PV panels are usually positioned at an angle to the horizontal plane, the energy input to the system must be calculated from the data. The approach here follows Duffie and Beckman (1991) and Lorenzo (1989) where further details can be found.

The block diagram of a standard calculation to determine the daily radiation on an inclined surface facing south is shown in Fig. 2.8. The calculation proceeds in three steps. In the first step, the data for the site are used to determine the separate diffuse and beam contributions to the global irradiation on the horizontal plane. This is done by using the extraterrestrial daily irradiation B_0 as a reference, and calculating the ratio $K_T = G/B_0$, known as the *clearness index*. K_T describes the (average) attenuation of solar radiation by the atmosphere at a given site during a given month. In the evaluation of B_0, the variation of the extraterrestrial irradiance on account of the eccentricity of the Earth's orbit (approximately $\pm 3\%$) is usually also taken into account.

In the second step, the diffuse irradiation is obtained using the empirical rule that the diffuse fraction D/G of the global radiation is a universal function of the clearness index K_T. Since $B = G - D$, this procedure determines both the diffuse and beam irradiation on the horizontal plane.

In the third step, the appropriate angular dependences of each component are used to determine the diffuse and beam irradiation on the inclined surface. With allowance for the reflectivity of the surrounding area, the albedo can also be determined. The total daily irradiation on the inclined surface is then obtained by adding the three contributions.

As we shall further discuss in Chapter 4, the daily irradiation values on an inclined panel for each month of the year play an important part in the design of a photovoltaic system.

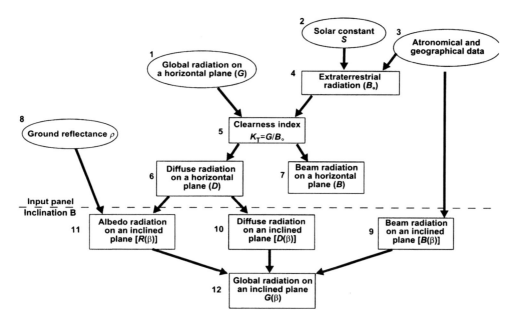

Fig. 2.8 Calculation of global radiation on an inclined plane. Oval boxes represent input data, and rectangular boxes indicate steps in the calculations

In more detail now, the calculation proceeds as follows. The inputs to the calculation are the global daily irradiation G for a day near the middle of each month of the year [block 1], the solar constant S and its variation due to the eccentricity of the Earth's orbit [block 2], the site's geographical latitude and the solar declination δ for the days of the year in question [block 3]. During the calculation, one also has to input the panel inclination β and the reflectivity ρ of the surrounding area [block 8]. Some typical values of surface reflectivity are given in Table 2.1.

[4] The radiation B_o received, over one day, by a unit horizontal area outside the Earth's atmosphere is calculated using the expression

$$B_o = \frac{24}{\pi} S\{1 + 0.33\cos(2\pi d_n/365)\}(\cos\phi\cos\delta\sin\omega_s + \omega_s\sin\phi\sin\delta)$$

where d_n is the number of the day in the year (1 on 1st January and 365 on 31st December).

[5] The clearness index K_T is now calculated for each month of the year from

$$K_T = G/B_o$$

[6, 7] Various empirical formulas are available for the calculation of the diffuse radiation. We shall adopt here the formula due to Page (1961) which is very simple and accurate enough for the present purposes:

$$D/G = 1 - 1.13\,K_T$$

Table 2.1 Typical reflectivity values for some ground covers. (Note: This table is presented for illustration only, as a considerable range of values is encountered for each of the materials. For details, the student should consult the bibliography of this chapter)

Ground cover	Reflectivity
Dry bare ground	0.2
Dry grassland	0.3
Desert sand	0.4
Snow	0.5–0.8

The beam irradiation is then calculated from

$$B = G - D$$

[9] The beam radiation $B(\beta)$ on a south-facing panel inclined at an angle β to the horizontal surface is now given by

$$B(\beta) = B\frac{\cos(\phi - \beta)\cos\delta\sin\omega_o + \omega_o\sin(\phi - \beta)\sin\delta}{\cos\phi\cos\delta\sin\omega_s + \omega_s\sin\phi\sin\delta} \quad (2.6)$$

Here,

$$\omega_0 = \min\{\omega_s, \omega'_s\}$$

where ω_s is the sunrise angle above the horizon given by Eq. (2.5) and

$$\omega'_s = \cos^{-1}\{-\tan(\phi - \beta)\tan\delta\}$$

is the sunrise angle above a plane inclined at angle β to the horizontal.
[10] Assuming that the diffuse radiation is distributed isotropically over the sky dome, the diffuse radiation on the inclined surface is given by

$$D(\beta) = \frac{1}{2}(1 + \cos\beta)D \quad (2.7)$$

[8, 11] The albedo radiation. The irradiance reflected from the ground is generally small, and a simple isotropic model is usually sufficient. This gives the result

$$R(\beta) = \frac{1}{2}(1 - \cos\beta)\rho G \quad (2.8)$$

The albedo radiation takes on special importance for photovoltaic modules (called *bifacial*) which can utilise energy incident both from the front and from the rear of the panel.
[12] We can now find the total global irradiation $G(\beta)$ on the inclined surface as a sum of the all the contributions (6)–(8):

$$G(\beta) = B(\beta) + D(\beta) + R(\beta)$$

As an illustration, Table 2.2 shows the data for different panel orientations obtained for Barcelona on the Mediterranean coast of Spain. These data will be used in Chapter 4 to discuss the detailed design of a PV system. Presently, we shall consider a few general rules regarding the orientation of the PV modules to capture the optimum amount of solar radiation.

To this end, Fig. 2.9 shows the mean daily irradiation in Barcelona during different parts of the year (summer, winter, and yearly average) as a function of

Table 2.2 Daily irradiation in Barcelona (in kWh/m^2) for a typical day in every month as a function of the panel inclination in degrees

Angle	Jan	Feb	Mar	Apr	May	Jun	Jul	Aug	Sep	Oct	Nov	Dec	Annual
0	2.25	2.92	3.88	4.98	5.66	6.38	6.67	5.75	4.20	3.18	2.23	1.70	4.15
5	2.56	3.19	4.08	5.09	5.69	6.36	6.68	5.84	4.46	3.42	2.49	1.93	4.32
10	2.86	3.44	4.26	5.18	5.70	6.32	6.66	5.90	4.61	3.64	2.74	2.16	4.46
15	3.14	3.67	4.42	5.25	5.67	6.25	6.61	5.94	4.73	3.84	2.98	2.37	4.57
20	3.40	3.87	4.55	5.28	5.62	6.15	6.52	5.94	4.82	4.03	3.19	2.57	4.66
25	3.65	4.05	4.66	5.29	5.54	6.03	6.41	5.91	4.89	4.18	3.39	2.75	4.73
30	3.86	4.21	4.73	5.26	5.44	5.88	6.26	5.85	4.93	4.31	3.57	2.92	4.77
35	4.05	4.34	4.78	5.21	5.31	5.70	6.08	5.75	4.94	4.42	3.72	3.07	4.78
40	4.22	4.45	4.81	5.13	5.15	5.49	5.88	5.63	4.93	4.50	3.85	3.19	4.77
45	4.36	4.53	4.80	5.03	4.97	5.26	5.65	5.48	4.88	4.55	3.96	3.30	4.73
50	4.47	4.58	4.77	4.89	4.77	5.01	5.39	5.29	4.81	4.57	4.04	3.39	4.66
55	4.55	4.60	4.71	4.73	4.55	4.74	5.11	5.09	4.71	4.57	4.09	3.45	4.57
60	4.60	4.59	4.62	4.55	4.30	4.45	4.80	4.85	4.58	4.53	4.12	3.49	4.46
65	4.62	4.55	4.50	4.34	4.04	4.14	4.48	4.59	4.42	4.47	4.12	3.51	4.32
70	4.61	4.49	4.36	4.11	3.77	3.83	4.15	4.31	4.25	4.38	4.10	3.51	4.15
75	4.57	4.39	4.19	3.86	3.48	3.50	3.80	4.02	4.05	4.27	4.04	3.48	3.97
80	4.50	4.27	4.00	3.59	3.18	3.17	3.44	3.70	3.82	4.13	3.97	3.43	3.77
85	4.40	4.13	3.79	3.31	2.88	2.84	3.08	3.37	3.58	3.96	3.87	3.36	3.55
90	4.27	3.95	3.55	3.02	2.57	2.51	1.86	3.04	3.32	3.78	3.74	3.27	3.24

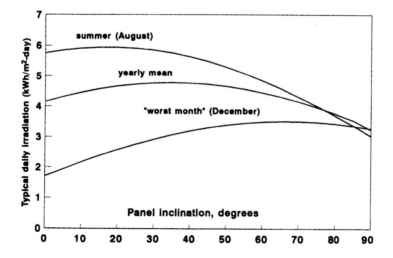

Fig. 2.9 Daily irradiation in Barcelona as a function of panel inclination

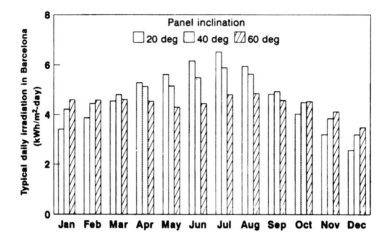

Fig. 2.10 Daily irradiation in Barcelona over the year for selected angles of panel inclination

the panel inclination. This will allow us to select the optimum inclination for some typical applications.

In some systems (for example, those connected to the grid), the paramount consideration is to collect the maximum energy over the year. It can be seen that the optimum panel angle which maximises the yearly average irradiation lies between 30° and 40°. In general, it is usually close to the latitude angle.

In many stand-alone systems which rely on energy storage by batteries, the principal consideration may not be the total energy received but the daily irradiation during the month with the least sunshine (here, December). In this instance, one should choose an angle between 60° and 70°.

Solar electricity is best suited for applications where peak consumption occurs during the summer (for example, crop irrigation). Here, a panel angle close to 20° would be the best option.

Figure 2.10 shows the typical daily irradiation in each month of the year for the three angles 20°, 40° and 60° which could be chosen in the above three applications.

SUMMARY OF THE CHAPTER

The nature of solar radiation on Earth was reviewed, and the principal components of the radiation were introduced. We have seen how the solar radiation, both in magnitude and structure, depends on the geographical location. The astronomical relationship between the Earth and the Sun was discussed, defining the principal angles which determine the apparent motion of the Sun. These results were used to evaluate the daily solar irradiation on an inclined surface which represents the energy input to a photovoltaic system.

We have seen that inclination close to the latitude angle of the site will maximise the radiation on an inclined panel over the whole year. In stand-alone systems, it is usual to choose a somewhat steeper angle to minimise storage requirements. For a summer peaking load, on the other hand, a more shallow angle is usually preferred.

BIBLIOGRAPHY AND REFERENCES

DUFFIE, J.A. and BECKMAN, W.A., *Solar Engineering of Thermal Processes (2nd edition)*, John Wiley & Sons, New York, 1991.

IQBAL, M., *An Introduction to Solar Radiation*, Academic, New York, 1983.

LOF, G. O. F., DUFFIE, F. A. and SMITH, C. O., *World Distribution of Solar Radiation*, University of Wisconsin Report No. 21, 1966.

LORENZO, E., Solar Radiation, *in:* Luque A., *Solar Cells and Optics for Photovoltaic Concentration*, Adam Hilger, Bristol, 1989, pp. 268–304.

PAGE, J. K., The estimation of monthly mean values of daily total short-wave radiation on vertical and inclined surfaces from sunshine records for latitudes 40 °N–40 °S, *in: Proc. United Nations on New Sources of Energy*, Vol. 4, 1961, pp. 378–390.

PALZ, W., ed. *European Solar Radiation Atlas*, Volumes 1 and 2, 2nd edn, Verlag TUV Rheinland, Cologne, 1984.

SELF-ASSESSMENT QUESTIONS

PART A

1. What is the approximate solar irradiance on a clear day with the Sun at zenith?
   (a) 1 W/cm^2; (b) 0.01 W/cm^2; (c) 0.1 W/cm^2.

2. What are the three components of solar radiation on an inclined panel?

3. What are the main physical processes that determine the attenuation of solar radiation in the atmosphere?

4. Which of the following statements are true?
   (a) The average daily irradiation over the year is maximum at the equator.
   (b) Regions with high latitude experience relatively high amounts of diffuse radiation.
   (c) The seasonal variation of daily solar irradiation is approximately independent of the latitude.

5. What are the main angular quantities that characterise the (a) yearly and (b) daily apparent motion of the Sun?

6. What is the maximum solar elevation at a site with with latitude 40 °N and when does it occur?

7. What measured solar radiation data are usually used to determine the energy input to the PV system?

8. Describe the principal steps of the calculation to determine the solar radiation on an inclined surface.

9. For a photovoltaic installation at latitude 30 °N, you have the option to position the panels at an angle 0°, 10°, 30°, 50°, or 90° to the horizontal plane. Which panel orientation would you choose to obtain the maximum energy: (a) over the year; (b) in the summer; (c) in the winter?

PART B

The mean daily solar radiation on the horizontal plane for a location at latitude 38.1 °N is given by the following Table:

	Jan	Feb	Mar	Apr	May	Jun	Jul	Aug	Sep	Oct	Nov	Dec
$G(\text{kWh/m}^2)$	2.62	3.11	4.75	5.51	6.66	6.92	7.34	6.33	5.02	3.84	2.46	2.30

(i) Determine the mean daily solar radiation B_0 outside the Earth's atmosphere for each month of the year. Hence, find the values of the clearness index K_T.
(ii) Determine the daily global radiation for angles of incidence $\beta = 10°$, $20°$, ..., $90°$. assuming ground reflectivity 0.2.
(iii) What angle of inclination would you recommend to maximise input to PV system:
(a) during the worst month of the year;
(b) to obtain the highest energy incident on the inclined panels over the whole year;
(c) to obtain the highest energy during the summer months?

Answers

Part A

1. (c).
2. Direct (beam), diffuse and surface albedo.
3. Scattering and absorption.
4. (a) F; (b) T; (c) F (see end of section 1).
5. (a) Solar declination.
   (b) Hour angle or azimuth, and the solar elevation or zenith angle.
6. $90° - (40° - 23.5°) = 73.5°$; at solar noon on the day of summer equinox (21/22 June).
7. Global radiation on a horizontal surface.
8. See Fig. 2.8.
9. (a) 30°; (b) 10°; (c) 50°.

Part B

(i), (ii) See Table 2.3.

(iii) 60° to maximise daily radiation in November;
30° to maximise yearly radiation;
10–20° to maximise daily radiation in August.

Table 2.3 Solution to S.A.Q. Part B (i) and (ii)

| | B_o (kWh) | K_T | Global radiation on an inclined plane (kWh) | | | | | | | | | |
			0	10	20	30	40	50	60	70	80	90
Jan	4.55	0.58	2.62	3.23	3.75	4.19	4.52	4.74	4.83	4.80	4.65	4.37
Feb	6.01	0.52	3.11	3.55	3.91	4.18	4.35	4.41	4.38	4.24	4.00	3.66
Mar	7.93	0.60	4.75	5.19	5.50	5.69	5.74	5.66	5.44	5.09	4.63	4.06
Apr	9.77	0.56	5.51	5.70	5.77	5.71	5.53	5.24	4.83	4.33	3.75	3.11
May	11.05	0.60	6.66	6.67	6.55	6.30	5.93	5.44	4.86	4.20	3.48	2.75
Jun	11.57	0.60	6.92	6.83	6.61	6.28	5.83	5.27	4.64	3.95	3.22	2.51
Jul	11.33	0.65	7.34	7.28	7.08	6.75	6.28	5.70	5.01	4.26	3.46	2.67
Aug	10.33	0.61	6.33	6.45	6.44	6.29	6.01	5.60	5.08	4.46	3.77	3.04
Sep	8.73	0.58	5.02	5.33	5.52	5.58	5.52	5.33	5.02	4.60	4.09	3.49
Oct	6.73	0.57	3.84	4.32	4.70	4.97	5.13	5.15	5.06	4.84	4.51	4.08
Nov	5.00	0.49	2.46	2.89	3.25	3.54	3.74	3.85	3.88	3.80	3.64	3.39
Dec	4.17	0.55	2.30	2.86	3.36	3.77	4.09	4.31	4.41	4.40	4.28	4.05
Annual average			4.74	5.02	5.20	5.27	5.22	5.06	4.79	4.41	3.96	3.43

3

Solar Cells

AIMS

The aims of this unit are to explain how solar cells work and how they are manufactured.

OBJECTIVES

When you complete this unit you should be able to:

1. analyse the structure of a solar cell,
2. explain the power output from a cell in terms of the incident energy flux and the electronic structure of the semiconductor,
3. evaluate the cell performance by using its current–voltage characteristic,
4. assess the operation of practical devices and limits imposed on their performance,
5. identify the technological steps which are used in the manufacture of solar cells.

NOTATION AND UNITS

Symbol		SI unit	Other unit
A	Surface area of the cell	m^2	
c	Speed of light in vacuum	m/s	
E_c	Energy of the bottom of the conduction band	J	eV
E_v	Energy of the top of valence band	J	eV
E_g	Energy gap	J	eV
$E_{ph}(\lambda)$	Photon energy	J	eV
FF	Fill factor		

Written by A. Cuevas (Silicon solar cell technology), R. Hill (Thin-film solar cells) and T. Markvart.

(Contd)

G	Irradiance	W/m^2	
h	Planck constant	Js	
I	Electric current	A	
I_{sc}	Short-circuit current	A	
I_ℓ	Light-generated current	A	
I_m	Current at maximum power	A	
I_o	Dark saturation current	A	
J	Electric current density	A/m^2	
k	Boltzmann constant	J/K	eV/K
m	Nonideality factor		
$\mathcal{N}$	Photon flux (density)	$m^{-2}s^{-1}$	
P_{max}	Power produced by the cell at the maximum power point	W	
q	Electron charge	C	
R_s	Series resistance	Ω	
T	Absolute temperature	K	
V	Voltage	V	
V_{OC}	Open-circuit voltage	V	
V_m	Voltage at maximum power	V	
η	Cell efficiency		
λ	Wavelength of light	m	
	Spectral flux density	$m^{-2}s^{-1}eV^{-1}$	
		or	
		$m^{-2}s^{-1}\mu m^{-1}$	

Unit conversion factors
$1\,eV = 1.602 \times 10^{-19}\,J$

Values of physical constants
$k = 1.38 \times 10^{-23}\,J/K = 86.3 \times 10^{-3}\,eV/K$
$q = 1.602 \times 10^{-19}\,C$
$c = 2.998 \times 10^{8}\,m/s$
$h = 6.626 \times 10^{-34}\,Js$

3.1 INTRODUCTION

In this chapter, you will learn how solar cells work and how they are manufactured. A broad overview of the available solar cells will be given in section 3.2. You will then be shown in section 3.3 how solar cells work. Starting with the electronic properties of semiconductors we shall examine the overall structure of the cell and explain how it produces electricity. Various practical devices will be discussed, together with power losses which occur during their operation.

The devices that you are most likely to encounter—the silicon and thin-film solar cells—will then be analysed in detail in sections 3.4 and 3.5. You will be shown how they are manufactured, from raw materials to the final device. Emphasis will be placed not only on the electrical properties of the device, but also on the economic aspects of the technology.

3.2 WHAT ARE SOLAR CELLS?

Solar cells represent the fundamental power conversion unit of a photovoltaic system. They are made from semiconductors, and have much in common with other solid-state electronic devices, such as diodes, transistors and integrated circuits. For practical operation, solar cells are usually assembled into modules.

Many different solar cells are now available on the market, and yet more are under development (Fig. 3.1). The range of solar cells spans different materials and different structures in the quest to extract maximum power from the device while keeping the cost to a minimum. Devices with efficiency exceeding 30% have been demonstrated in the laboratory. The efficiency of commercial devices, however, is usually less than half this value.

Crystalline silicon cells hold the largest part of the market. To reduce the cost, these cells are now often made from multicrystalline material, rather than from the more expensive single crystals. Crystalline silicon cell technology is well established. The modules have a long lifetime (20 years or more) and their best production efficiency is approaching 18%.

Cheaper (but also less efficient) types of silicon cells, made in the form of amorphous thin films, are used to power a variety of consumer products. You will be familiar with the solar-powered watches and calculators, but larger amorphous silicon solar modules are also available.

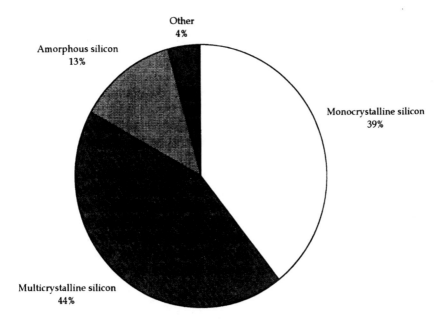

Fig. 3.1 Market share of the principal photovoltaic technologies (1998). 'Other' technologies include crystalline silicon cells for concentrator systems, cells based on ribbon silicon, cadmium telluride and silicon on low-cost substrates (Source: Paul Maycock, PV News).

A variety of compound semiconductors can also be used to manufacture thin-film cells (for example, cadmium telluride or copper indium diselenide). These modules are now beginning to appear on the market and hold the promise of combining low cost with acceptable conversion efficiencies.

A particular class of high-efficiency solar cells from single crystal silicon or compound semiconductors (for example, gallium arsenide or indium phosphide) are used in specialised applications, such as to power satellites or in systems which operate under high-intensity concentrated sunlight. The operation and applications of these devices will be reviewed in Chapters 5 and 7.

Photovoltaic materials are not restricted to semiconductors. Solar cells are now available which convert light to electricity by organic molecules, with best conversion efficiency exceeding 10%. The principles of this novel type of solar energy conversion are summarised briefly in Section 3.3, and discussed in detail in Section 7.3.

Summary

Various types of solar cells have been introduced: crystalline and amorphous silicon cells, compound thin-film devices, and high-efficiency cells for specialised applications.

3.3 HOW SOLAR CELLS WORK

3.3.1 Introduction

The solar cell operation is based on the ability of semiconductors to convert sunlight directly into electricity by exploiting the *photovoltaic effect*. In the conversion process, the incident energy of light creates mobile charged particles in the semiconductor which are then separated by the device structure and produce electrical current.

The electronic structure of semiconductors will be reviewed in section 3.3.2. You will learn here about the characteristic distribution of electron energies within the semiconductor, and how the electrical properties of semiconductors can be controlled by the addition of impurities. We shall introduce an essential part of the solar cell, the semiconductor junction, and show how the illumination creates mobile charged particles, electrons and holes. The electrical characteristics of the solar cell will then be obtained by analysing the charge currents across the junction.

In section 3.3.4, various device structures will be reviewed and we show how the device design aims to minimise the losses of power which occur during the cell operation. We shall also examine how the operation of the cell is affected by practical operating conditions, particularly by variable temperature and irradiance.

3.3.2 Electronic structure of semiconductors

3.3.2.1 Band structure, doping

The principles of semiconductor physics are best illustrated by the example of silicon, a group 4 elemental semiconductor. The silicon crystal forms the so-called *diamond lattice* where each atom has four nearest neighbours at the vertices of a tetrahedron. The four-fold tetrahedral coordination is the result of the bonding arrangement which uses the four outer (valence) electrons of each silicon atom (Fig. 3.2). Each bond contains two electrons, and you can easily see that all the valence electrons are taken up by the bonds. Most other industrially important semiconductors crystallise in closely related lattices, and have a similar arrangement of the bonding orbitals.

This crystal structure has a profound effect on the electronic and optical properties of the semiconductor.

According to the quantum theory, the energy of an electron in the crystal must fall within well-defined *bands*. The energies of valence orbitals which form bonds between the atoms represent just such a band of states, the *valence band*. The next higher band is the *conduction band* which is separated from the valence band by the *energy gap*, or *bandgap*. The width of the bandgap $E_c - E_v$ is a very important characteristic of the semiconductor and is usually denoted by E_g. Table 3.1 gives the bandgaps of the most important semiconductors for solar-cell applications.

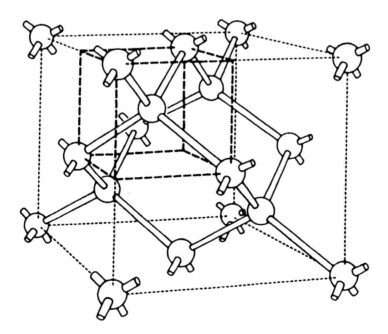

Fig. 3.2 The diamond lattice

Table 3.1 Energy gaps of principal semiconductors for photovoltaic applications (gap values given at room temperature)

Material	Energy gap (eV)	Type of gap
crystalline Si	1.12	indirect
amorphous Si	~1.75	direct
$CuInSe_2$	1.05	direct
CdTe	1.45	direct
GaAs	1.42	direct
InP	1.34	direct

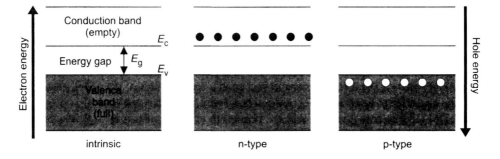

Fig. 3.3 Band diagram and electron–hole distribution in semiconductors

A pure semiconductor (which is called *intrinsic*) contains just the right number of electrons to fill the valence band, and the conduction band is therefore empty (Fig. 3.3). Electrons in the full valence band cannot move—just as, for example, marbles in a full box with a lid on top. For practical purposes, a pure semiconductor is therefore an insulator.

Semiconductors can only conduct electricity if carriers are introduced into the conduction band or removed from the valence band. One way of doing this is by alloying the semiconductor with an impurity. This process is called *doping*. As we shall see, doping makes it possible to exert a great deal of control over the electronic properties of a semiconductor, and lies at the heart of the manufacturing process of all semiconductor devices.

Suppose that some group 5 impurity atoms (for example, phosphorus) are added to the silicon melt from which the crystal is grown. Four of the five outer electrons are used to fill the valence band and the one extra electron from each impurity atom is therefore promoted to the conduction band (Fig. 3.3). For this reason, these impurity atoms are called *donors*. The electrons in the conduction band are mobile, and the crystal becomes a conductor. Since the current is carried by negatively charged electrons, this type of semiconductor is called *n-type*.

A similar situation occurs when silicon is doped with group 3 impurity atoms (for example, boron) which are called *acceptors*. Since four electrons per atom are needed to fill the valence band completely, this doping creates electron deficiency in this band (Fig. 3.3). The missing electrons—called *holes*—behave as positively charged particles which are mobile, and carry current. A semiconductor where the electric current is carried predominantly by holes is called *p-type*.

The prevailing charge carriers in a given semiconductor are called *majority carriers*. Examples of majority carriers are electrons in an n-type semiconductor and holes in the p-type. The opposite type of carriers whose concentration is generally much lower, are called *minority carriers*.

3.3.2.2 Semiconductor junctions

The operation of solar cells is based on the formation of a *junction*. Various examples of junctions are shown in Fig. 3.4. Perhaps the simplest is the *p-n junction*, an interface between the n and p regions of one semiconductor. A layer of intrinsic material is sometimes incorporated between the n- and p-type regions, resulting in a wider transition zone. In contrast with these *homojunctions*, a *heterojunction* is formed by two different semiconductors—note the difference in the bandgaps on the two sides of the junction.

An interface between a metal and a semiconductor may also form a junction, called the *Schottky barrier*. In general, the properties of metal contacts with a semiconductor depend on the actual materials in question. For each semiconductor, some metals form a Schottky barrier but some form an *ohmic contact* where the barrier is absent. These contacts are used to extract electrical current from the device.

The important feature of all junctions is that they contain a strong electric field. To illustrate how this field comes about, let us imagine the hypothetical situation where the p-n junction is formed by joining together two pieces of semiconductor, one p-type and the other n-type (although this manner of junction formation is not normally employed in practice, it is a convenient means to demonstrate the relevant principles). In separation, there is electron surplus in the n-type material and hole surplus in the p-type. When the two pieces are brought into contact, electrons from the n region near the interface diffuse into the p side, leaving behind a layer which is positively charged by the donors. Similarly, holes diffuse in the opposite direction, leaving behind a negatively charged layer stripped of holes. The resulting junction region then contains practically no mobile charge carriers (Fig. 3.5), and the fixed charges of the dopant atoms create a potential barrier acting against a further flow of electrons and holes. Note that the electric field in the junction pulls the electrons and holes in opposite directions.

Figure 3.6 shows the band diagram of a p-n junction diode in equilibrium, and when external voltage is connected to the diode. Without bias, of course, there is no current through the junction. We may imagine this zero net current as consisting of two very small opposite currents I_o and $-I_o$ which remain from the current flow prior to the junction formation. These currents are very small indeed, corresponding to a current density of the order of $10^{-14} A/cm^2$ in a good silicon diode.

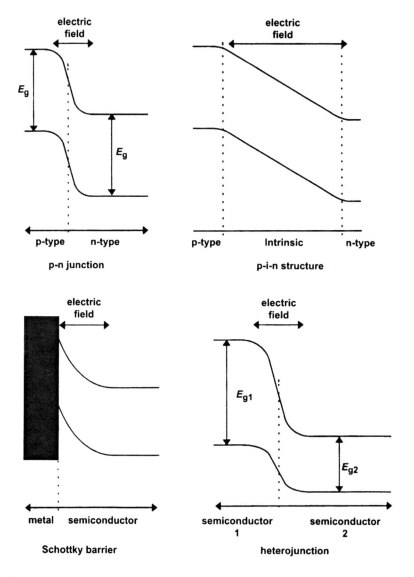

Fig. 3.4 Band diagram of semiconductor junctions

This current balance is altered considerably when voltage is applied to the junction. A forward bias, i.e. positive voltage applied to the p side, reduces the height of the potential barrier. This, in turn, increases dramatically the current through the diode.

Under reverse bias, on the other hand, the barrier is increased. This has a much smaller effect on the device, and produces only the tiny current I_o (the *dark saturation current*)—which is much smaller than the current under forward bias. The junction therefore acts as a rectifier, or diode.

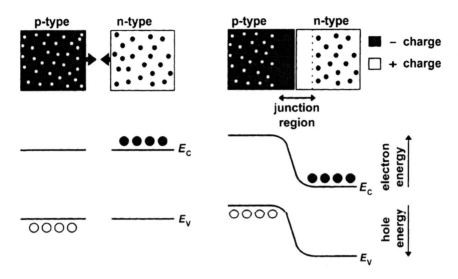

Fig. 3.5 Diagram of p-n junction formation and the resulting band structure

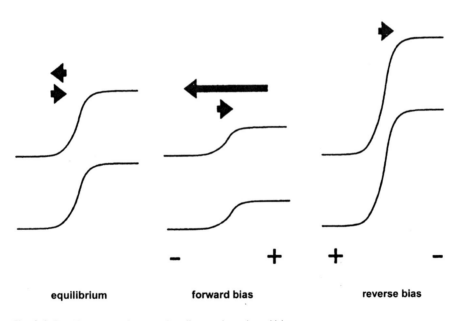

Fig. 3.6 Electric currents in a p-n junction under external bias

In mathematical terms, the I–V characteristic of a diode is given by the Shockley equation

$$I = I_0 \left[\exp\left(\frac{qV}{kT}\right) - 1 \right] \tag{3.1}$$

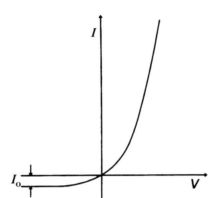

Fig. 3.7 The diode *I—V* characteristic

where I is the current, V is the voltage, k is the Boltzmann constant, q is the magnitude of the electron charge, and T is the absolute temperature. The *I–V* characteristic (3.1) is shown in Fig. 3.7.

3.3.2.3 Light absorption by a semiconductor

Photovoltaic energy conversion relies on the quantum nature of light whereby we perceive light as a flux of particles—*photons*—which carry the energy

$$E_{ph}(\lambda) = \frac{h\,c}{\lambda} \qquad (3.2)$$

where h is the Planck constant, c is the speed of light, and λ is the wavelength of light. On a clear day, about 4.4×10^{17} photons strike a square centimetre of the Earth's surface every second.

Only some of these photons—those with energy in excess of the bandgap—can be converted into electricity by the solar cell. When such a photon enters the semiconductor, it may be absorbed and promote an electron from the valence to the conduction band (Fig. 3.8). Since a hole is left behind in the valence band, the absorption process generates *electron–hole pairs*.

Each semiconductor is therefore restricted to converting only a part of the solar spectrum (Fig. 3.9). Using equation (3.2), the spectrum has been plotted here in terms of the incident photon flux as a function of photon energy. The shaded area represents the photon flux that can be converted by a silicon cell—about two thirds of the total flux.

The nature of the absorption process also indicates how a part of the incident photon energy is lost in the event. Indeed, it is seen that practically all the generated electron–hole pairs have energy in excess of the bandgap. Immediately after their creation, the electron and hole decay to states near the edges of their respective bands. The excess energy is lost as heat and cannot be converted into useful power. This represents one of the fundamental loss mechanisms in a solar cell.

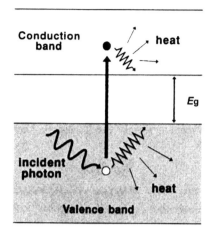

Fig. 3.8 The generation of electron–hole pairs by light

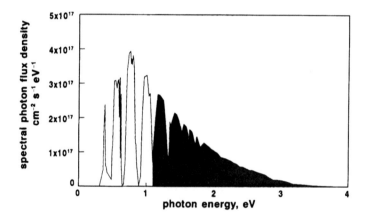

Fig. 3.9 Photon flux utilised by a silicon solar cell

We can make some rough estimates of the magnitude of electrical power that can be produced. To this end, let us interpret the light-induced electron traffic across the bandgap as electron current, called the *generation current*. We shall see shortly (section 3.3.3) that a solar cell can indeed transform this fictitious current into real electric current across the device. Neglecting losses, each photon then contributes one electron charge to the generation current. The electric current is then equal to

$$I_\ell = q \mathcal{N} A \tag{3.3}$$

where $\mathcal{N}$ is the number of photons in the highlighted area of the spectrum, and A is the surface area of the semiconductor that is exposed to light. For example, the current density $J_\ell = I_\ell / A$ that corresponds to the terrestrial spectrum is about $1.6 \times 10^{-19} \times 4.4 \times 10^{17} = 70 \, \text{mA/cm}^2$. Of this, a silicon solar cell can convert at most $44 \, \text{mA/cm}^2$.

What voltage can a solar cell generate? One can obtain an upper bound by a simple electrostatic argument. As we shall see, the electric power is produced by separating the light-generated electrons and holes to the terminals of the device. This separation can only proceed if the electrostatic energy of the charges after separation (qV where V is the voltage at the terminals) does not exceed the pair energy in the semiconductor, equal to the bandgap. This sets an upper limit on the voltage in the form

$$V = E_g/q \tag{3.4}$$

The *maximum voltage* in volts is thus numerically equal to the bandgap of the semiconductor in electronvolts. Although the actual voltage achieved in practice is considerably less than the theoretical limit, the trend expressed by (3.4) that wide-bandgap semiconductors produce higher voltage generally holds true.

We have so far assumed that all the photons with energy in excess of the bandgap are absorbed. Indeed, many semiconductors are good light absorbers, and absorb all the above-bandgap light in a layer of few micrometres thick. They are called *direct-bandgap semiconductors*. In others—*indirect-gap semiconductors* which include also crystalline silicon—the absorption process is more complicated. A quantum of lattice vibrations must participate in the conversion of a photon into an electron–hole pair to conserve momentum which hinders the process and decreases the capability of the semiconductor to absorb light. This phenomenon is illustrated in Fig. 3.10. Note that several hundred micrometres of silicon are necessary to absorb all the above-bandgap light but few micrometres of a direct-gap material (for example, GaAs) are sufficient for this purpose.

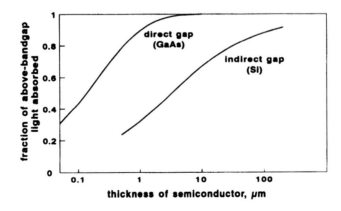

Fig. 3.10 Optical absorption properties of direct- and indirect-gap semiconductors

3.3.3 The solar cell

Figure 3.11 shows the diagram of a silicon cell, the typical solar cell in use today. The electrical current generated in the semiconductor is extracted by contacts to the front and rear of the cell. The top contact structure which must allow light to pass through is made in the form of widely-spaced thin metal strips (usually called *fingers*) that supply current to a larger bus bar. The cell is covered with a thin layer of dielectric material—the *antireflection coating* or ARC—to minimise light reflection from the top surface.

Figure 3.12(a) shows the band diagram of the semiconductor section under illumination. Light generates electron–hole pairs on both sides of the junction, in the n-type emitter and in the p-type base. The generated minority carriers—electrons from the base and holes from the emitter—then diffuse to the junction and are swept away by the electric field, thus producing electric current across the device. Note how the electric currents of the electrons and holes reinforce each other since these particles carry opposite charges. The p-n junction therefore separates the carriers with opposite charge, and transforms the generation current I_ℓ between the bands into an electric current across the p-n junction.

The I–V characteristic of a solar cell can be obtained by drawing an equivalent circuit of the device (Fig. 3.13). The generation of current I_ℓ by light is represented by a current generator in parallel with a diode which represents the p-n junction. The output current I is then equal to the difference between the light-generated current I_ℓ and the diode current I_D. Equation (3.1) then gives

$$I = I_\ell - I_o \left[\exp\left(\frac{qV}{kT}\right) - 1 \right] \tag{3.5}$$

Note that, under open circuit when $I = 0$, all the light-generated current passes through the diode. Under short circuit ($V = 0$) on the other hand, all this current passes through the external load. The I–V characteristic (3.5) and its relationship to the diode characteristic are shown in Fig. 3.13.

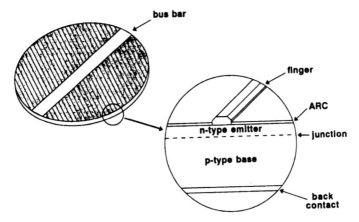

Fig. 3.11 The silicon solar cell

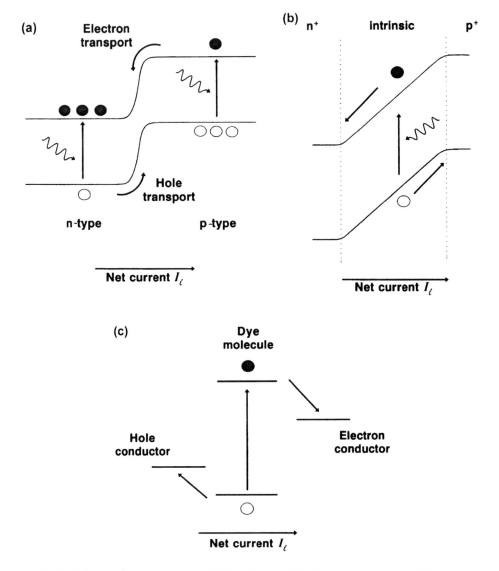

Fig. 3.12 Schematic representation of different types of photovoltaic converters. (a) Currents in a p-n junction under illumination (applicable, for example, to crystalline silicon or gallium arsenide solar cells); (b) the band diagram and operation of a p-i-n amorphous silicon solar cell; (c) energy conversion by a dye-sensitised photochemical solar cell

The I–V characteristic contains several important points. One is the short-circuit current I_{sc} which, as we noted, is simply the light-generated current I_ℓ. The second is the open-circuit voltage V_{oc} obtained by setting $I = 0$:

$$V_{\mathrm{oc}} = \frac{kT}{q} \ln\left(\frac{I_\ell}{I_0} + 1\right) \tag{3.6}$$

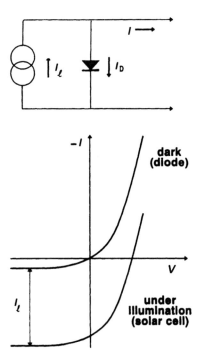

Fig. 3.13 The equivalent circuit and I–V characteristic of a solar cell compared to a diode

It is worthwhile to examine this equation in more detail. Both I_ℓ and I_o depend on the structure of the device. However, it is the value of I_o—which can vary by many orders of magnitude, depending on the device geometry and processing—that determines the open circuit voltage in practical devices.

No power is generated under short or open circuit. The maximum power P_{max} produced by the device is reached at a point on the characteristic where the product IV is maximum. This is shown graphically in Fig. 3.14 where the position of the *maximum power point* represents the largest area of the rectangle shown. One usually defines the fill factor FF by

$$P_{max} = V_m I_m = FF V_{oc} I_{sc} \tag{3.7}$$

where V_m and I_m are the voltage and current at the maximum power point.

The *efficiency* η of a solar cell is defined as the power P_{max} produced by the cell at the maximum power point under standard test conditions, divided by the power of the radiation incident upon it. Most frequent conditions are: irradiance $100\,\text{mW/cm}^2$, standard reference AM1.5 spectrum, and temperature $25\,°\text{C}$. The use of this standard irradiance value is particularly convenient since the cell efficiency in percent is then numerically equal to the power output from the cell in mW/cm^2. Other test conditions are sometimes adopted for cells which operate in a different environment, for example, cells which power satellites and operate under AM0 spectrum.

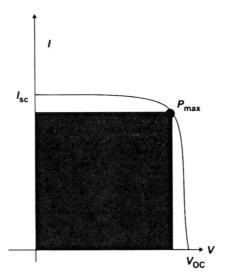

Fig. 3.14 The *I–V* characteristic of a solar cell with the maximum power point

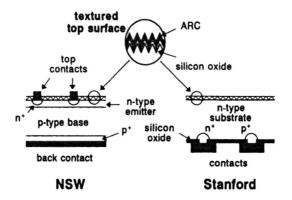

Fig. 3.15 The structure of high-efficiency silicon solar cells

The *I–V* characteristic (3.5) which we have derived for a simplified device describes, in fact, rather well the operation of solar cells in practice if the parameters make allowance for the losses which take place in practical devices. The typical production silicon cell (Fig. 3.11) has already been mentioned. Some modern types of research silicon cells are shown in Fig. 3.15. The *passivated emitter solar cell* (PESC) structure has been developed at the University of New South Wales in Australia for operation under ordinary sunlight. The *point-contact cell* of Stanford University, USA, has been designed for optimum operation under concentrated sunlight.

Gallium arsenide solar cells (Fig. 3.16) are, because of their high cost, usually intended for operation on satellites or in concentration systems. Gallium

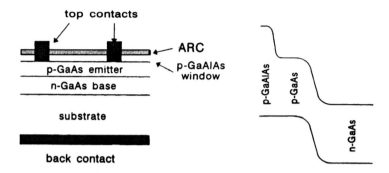

Fig. 3.16 The structure and band diagram of gallium arsenide solar cells

arsenide is a direct-gap semiconductor and most photons of light are absorbed in the top emitter layer. The top 'window' layer prevents these carriers diffusing to the top surface and being lost by surface recombination (see also section 3.3.4).

Most thin-film solar cells are made from amorphous or polycrystalline semiconductors with low diffusion constant for electrons and holes. To aid the carrier transport, these cells usually incorporate a lightly doped or intrinsic layer as part of the junction where most of the light is absorbed. Electrons and holes which are created in this layer are then pulled apart by the electric field immediately after their creation, eliminating carrier diffusion to the junction (Fig. 3.12(b)). The operation and manufacture of these cells will be discussed more fully in section 3.5.

Semiconductors need not form the key element of solar cell operation. Working solar cells have now been manufactured where the charge separation step is mediated by a molecular dye (Fig. 3.12(c)). In these devices, the dye layer covers a nanocrystalline titanium oxide electrode which acts as a receptor for electrons from the photoexcited dye molecules. The nanocrystalline structure of the titanium oxide particles assists efficient light absorption even by a very thin dye layer, which is probably monomolecular. The positive electrode is formed by a hole-carrying redox electrolyte. The operation of these cells is discussed in more detail in section 7.4.

3.3.4 Power losses in solar cells

Fundamental losses. As we have seen in section 3.3.2.3, carrier generation in the semiconductor by light involves considerable dissipation of the generated carrier energy into heat. In addition, a considerable part of the solar spectrum is not utilised because of the inability of a semiconductor to absorb the below-bandgap light.

Can these losses be reduced? Yes, but not with a simple structure that we have in mind at the moment. Such a device is called a *tandem cell* (Fig. 3.17), and represents a stack of several cells, each operating according to the principles that we have described. The top cell must be made of a high-bandgap

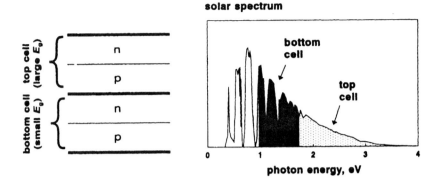

Fig. 3.17 The structure and spectral contributions of the tandem cell

semiconductor, and converts the short-wavelength radiation. The transmitted light is then converted by the bottom cell. This arrangement increases considerably the achievable efficiency. Laboratory devices operating at over 30% have been demonstrated. Most of these structures are at an experimental stage but some thin-film devices of this type may not take long to reach fruition, as we shall see in section 3.5.

Recombination. An opposite process to carrier generation is recombination when an electron–hole pair is annihilated. Recombination is most common at impurities or defects of the crystal structure, or at the surface of the semiconductor where energy levels may be introduced inside the energy gap. These levels act as stepping stones for the electrons to fall back into the valence band and recombine with holes (Fig. 3.18). An important site of recombination are also the ohmic metal contacts to the semiconductor.

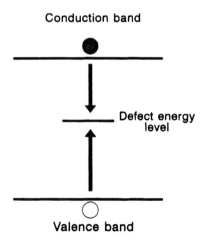

Fig. 3.18 Defect-assisted recombination of electron–hole pairs

What measures can one take to minimise the recombination losses? Surface recombination and recombination at contacts which are considerable in the conventional silicon cell (Fig. 3.11) can be reduced by adapting the device structure. Figures 3.15 and 3.16 show the implementation of such measures in high-efficiency silicon cells. The external surfaces of the semiconductor are here protected by a layer of passivating oxide to reduce surface recombination. The top layer of GaAlAs in the GaAs cell has a similar purpose. The contacts are surrounded by heavily-doped regions acting as 'minority-carrier mirrors' which impede the minority carriers from reaching the contacts and recombining.

Recombination reduces both the voltage and current output from the cell. The voltage losses reduce the limiting value (3.4) to the open-circuit voltage (3.6), restricted principally by the dark saturation current I_0. This current contains two contributions, each proportional to the number of recombination events—and therefore the volume—for each region on either side of the junction. This result is exploited in the Stanford point-contact silicon cell (Fig. 3.15) with a greatly reduced emitter volume. A thin silicon cell design which increases the voltage by reducing the base thickness, will be discussed in section 3.5.

The current losses can be grouped under the term of *collection efficiency*, the ratio between the number of carriers generated by light and the number that reaches the junction. Considerations of the collection efficiency affect the design of the solar cell. In crystalline materials, the transport properties are usually good, and carrier transport by simple diffusion is sufficiently effective. In amorphous and polycrystalline thin films, however, electric fields are needed to pull the carriers, as can be seen in the design of the amorphous silicon p-i-n cell (Fig. 3.12(b)).

Other losses to the current produced by the cell arise from light reflection from the top surface, shading of the cell by the top contacts, and incomplete absorption of light. The last feature is particularly significant for crystalline silicon cells since—as we have seen in section 3.3.2.3—silicon has poor light absorption properties. Various measures can be adopted to minimise these losses.

The top contact shading is of particular concern for cells which operate under concentrated, high-intensity sunlight. It is then advantageous to eliminate the shading altogether, as in the Stanford point-contact cell, by moving both contacts to the back of the cell.

The top-surface reflection is usually in the region of 10% for a cell covered with a single-layer ARC, and can be reduced further by the application of two or more layers. Another technique is *surface texturing*. As we shall see in section 3.4, the top surface of a silicon cell can be made in the form of small pyramids which give the reflected light another chance of absorption. When combined with a single-layer ARC, the reflectivity of the surface can be brought down to below 1%.

The absorption of light in silicon cells can be improved by making the back contact optically reflecting. When combined with a textured top surface, this geometry results in effective *light trapping* which provides a good counter-measure for the low absorptivity of silicon. In particular, as we shall see in section 3.5, it allows the manufacture of thin silicon cells.

Series resistance. The transmission of electric current produced by the solar cell involves ohmic losses. These can be grouped together and included as a resistance in the equivalent circuit (Fig. 3.19). It is seen that the series resistance affects the cell operation mainly by reducing the fill factor.

The *I–V* characteristic of a practical device is sometimes best approximated by a modified expression

$$I = I_\ell - I_o \left[\exp\left(\frac{qV + IR_s}{mkT} \right) - 1 \right] \tag{3.8}$$

which also includes the series resistance R_s. Here, m is an empirical *nonideality factor* whose value is usually close to unity. The light-generated current in (3.8) is taken as a measured parameter which takes into account all the current losses in the cell.

The losses which we have discussed in this section are summarised in Fig. 3.20. It is seen that the fundamental losses reduce the maximum theoretical efficiency of a silicon cell to about 48%. Additional voltage losses ($\sim$36%), current losses ($\sim$10%), and losses associated with the fill factor ($\sim$20%) then explain the efficiency of about 23% for the best silicon cell today. Higher efficiencies have been obtained under concentrated sunlight, in devices from

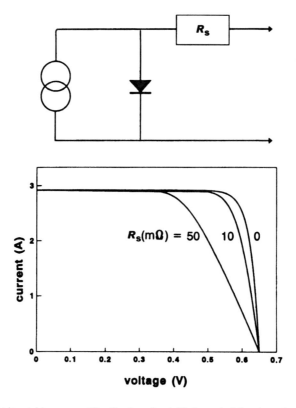

Fig. 3.19 The series resistance and its effect on the *I–V* characteristic of a solar cell

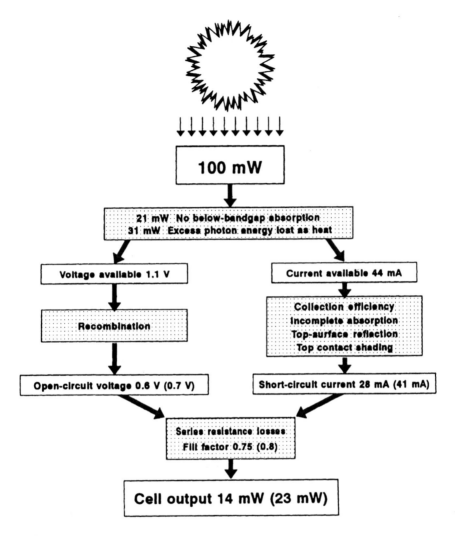

Fig. 3.20 The power losses in silicon solar cells. The figures are per square centimetre for production cells and (in brackets) for laboratory cells

other materials or in tandem structures. The typical production silicon cell has an efficiency of about 14%. However, new devices with efficiency approaching 18% are beginning to appear on the market.

3.3.5 Temperature and irradiance effects

In practical applications, solar cells do not operate under standard conditions. The two most important effects that must be allowed for are due to the variable temperature and irradiance.

Temperature. This has an important effect on the power output from the cell (Fig. 3.21). The most significant is the temperature dependence of the voltage which decreases with increasing temperature (*its temperature coefficient is negative*). The voltage decrease of a silicon cell is typically 2.3 mV per °C. The temperature variation of the current or the fill factor are less pronounced, and are usually neglected in the PV system design.

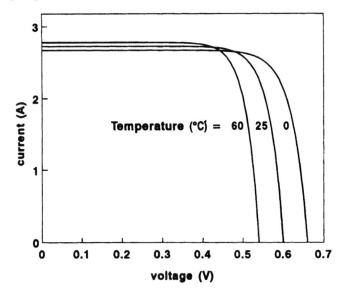

Fig. 3.21 Temperature dependence of the *I–V* characteristic of a solar cell

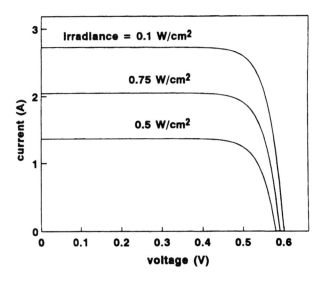

Fig. 3.22 Irradiance dependence of the *I–V* characteristic of a solar cell

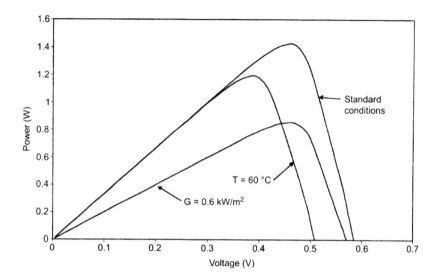

Fig. 3.23 Power produced by solar cell as a function of voltage V at the terminals

Effects of irradiance. The solar-cell characteristics under different levels of illumination are shown in Fig. 3.22. As we have seen, the light-generated current is proportional to the flux of photons with above-bandgap energy. Increasing the irradiance increases, in the same proportion, the photon flux which, in turn, generates a proportionately higher current. Therefore, the short-circuit current of a solar cell is directly proportional to the irradiance. The voltage variation, as can be seen from equation (3.6), is much smaller (it depends logarithmically on the irradiance), and is usually neglected in practical applications.

The power $P = IV$ produced by the solar cell as a function of voltage V at the cell terminals is depicted in Fig. 3.23. Also shown are the power characteristics at a lower irradiance and at elevated temperature. Although it is desirable to operate the cell at the maximum power point, this may not be easy to realise in practice. A simpler but less efficient solution is to operate the cell at a constant voltage below the voltage of the maximum power point. If the operating voltage remains in the linear part of the $I-V$ characteristic, temperature will have little effect on the power output of the solar cell. The power delivered to the load will therefore be proportional to the short-circuit current and thus also to the irradiance. This point will be taken up again in section 4.3.2. where we discuss the operation of the photovoltaic module.

Summary

The solar cell is a semiconductor device that converts the quantum flux of photons into electric current. When light is absorbed, it first creates electron–hole pairs. These mobile charges are then separated by the electric fields at the junction. The electrical output from the cell is described by the $I-V$ characteristic whose parameters can be linked to the material properties of the semiconductor. Various solar-cell structures have been discussed in relation to the principal power losses in a solar cell. In addition to the fundamental losses associated with light absorption, other losses, including recombination and losses dependent on the structure of the device, have been analysed in some detail.

3.4 SILICON SOLAR CELL TECHNOLOGY

3.4.1 Introduction

The technology based on the elemental semiconductor silicon, either single crystalline or multicrystalline, is the best-established photovolatic technology at the present time. It has many points in common with the microelectronics industry, and the benefits of those close links have helped to improve the performance of laboratory silicon solar cells to nearly 25% – a level not far from the theoretical maximum. While such high performance requires the use of high-quality materials as well as complex technology and equipment, the photovoltaic industry has developed remarkably simple fabrication processes capable of manufacturing large area solar cells at low cost. In particular, new technologies for the crystallisation and slicing of silicon have been developed, and simplified methods for doping the wafers and forming the metal contacts have been implemented. In 1998 the terrestrial photovoltaic market of about 130 MW_p (leaving an extra 20 MW_p of amorphous silicon aside, see Fig 3.1) was equally shared by the conventional single crystal and multicrystalline silicon technologies. The efficiency of commercial modules made with silicon solar cells lies in the range 10–15% (11–16% cell efficiencies), and there is a strong push towards developing advanced fabrication processes to increase the conversion efficiency. The reason for this is that the balance of system costs plays an important role in the final cost of photovoltaic electricity, and many of those costs (land, structures, etc) are practically the same for either low or high efficiency PV modules.

   An overview of the different steps of silicon solar cell fabrication follows. The conventional technology is described in some detail since it represents not only the past development but, to a large extent, the present state of the industrial production. New techniques that have already made a significant impact on silicon cell manufacture and those still in the research phase are also briefly described since they represent the likely evolution of the technology in the coming years. Four major stages need to be followed to make photovoltaic modules (the first one is not performed by the PV industry):

- from sand to pure silicon;
- from silicon feedstock to silicon crystals and wafers;
- from silicon wafers to solar cells;
- from cells to modules.

To understand the importance of the different stages, the breakdown of the cost of a PV module, which is about $4/W_p$ at present, should be considered. Roughly speaking, the cost is made up of three approximately equal contributions: 1/3 from the silicon material itself (in wafer form), 1/3 from the fabrication of the solar cells, and 1/3 from the encapsulation and construction of the module. Some authors give an even greater weight to the cost of the silicon wafers (50% of the total). It is clearly important to reduce the cost of the silicon but this alone is not sufficient to reduce the cost of PV electricity arbitrarily. Given the present cost structure, a reasonable strategy could be to tolerate an increase in the cost of the cell fabrication in exchange for a higher efficiency.

3.4.2 From sand to pure silicon

The supply of silicon is practically endless: although pure silicon does not occur in nature, 60% of the Earth's crust is sand, for the major part, quartzite or silicon dioxide (SiO_2). Silicon is produced in large amounts, more than 600 000 tonnes a year world-wide, to make special steel and alloys. This *metallurgical-grade silicon* (MG-Si) is obtained by reducing quartzite with coke (coal) in electric-arc furnaces. Its purity is only 98–99% – insufficient for electronic applications – but both the energy input (about 50 kWh/kg) and cost (about $2/kg) are relatively low.

The semiconductor industry purifies this metallurgical-grade silicon until the impurity concentration is less than 0.1 ppma (parts per million atomic). An alternative way of expressing the 0.1 ppma purity is as relative silicon concentration equal to 99.99999% – that is, seven nines. This defines the specification for *semiconductor-grade silicon*. Since 1 cm^3 of crystalline silicon contains 5×10^{22} atoms, this purity implies that the total number of foreign atoms must be less that $5 \times 10^{15}/\text{cm}^3$. The conventional purification process used by the semiconductor industry is described in Fig. 3.24. It starts with the chemical transformation of Si into a liquid compound (trichlorosilane, $SiHCl_3$, or silicon tetrachloride, $SiCl_4$, as an alternative) which allows the desired purity to be reached by multiple distillation. Finally, the trichlorosilane vapour is reduced with hydrogen to obtain the desired ultrapure elemental silicon in solid form. This final solidification step is an example of a type of technique generically called chemical vapour deposition (CVD): the reaction between the silicon compound and hydrogen takes place in the presence of a thin (0.5 cm diameter) silicon rod that is heated to a high temperature (about 1100 °C). As a product of the reaction, elemental silicon deposits onto the 2 m long rod, increasing its diameter to about 12.5 cm. The silicon rod is subsequently broken into chunks and packaged. This technique, which is known as the Siemens process, is very energy intensive (about 200 kWh/kg). The resulting material is not meant to meet any particular specification in terms of crystallographic structure and is commonly called *polysilicon* (or *poly-crystalline silicon*) to indicate that it is composed of an aggregate of micro-crystals. The present world production of semiconductor grade polysilicon is 20 000 tonnes and its price is approximately $50/kg.

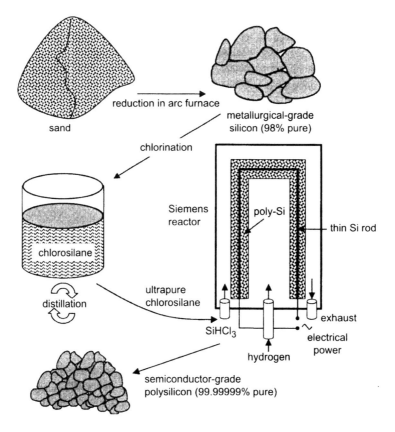

Fig. 3.24 Preparation of ultra-pure silicon from sand

It is worth mentioning that some types of *poly-crystalline silicon* are currently being investigated for thin-film silicon solar cells. Nevertheless, the specific techniques and deposition conditions needed to obtain poly-crystalline silicon films with good electronic properties are very different from the Siemens process discussed here (they are more akin to the techniques used to fabricate amorphous silicon solar cells, discussed later on in section 3.5.2). An added advantage of such a thin-film silicon approach is that it avoids the Siemens process and the consequent crystallisation and wafering steps discussed below.

Semiconductor-grade silicon is widely believed to be more pure than is required to make industrial solar cells. Although there is no universal definition of its purity, the so-called *solar-grade silicon* may contain up to about ten times more impurities (i.e. 1 ppma) and still permit reasonably efficient cells. Such a relaxation of the impurity contents should facilitate the development of lower cost purification procedures. Several techniques have already been investigated on a laboratory scale, although a significant effort is still needed to prove and develop a commercially feasible technology.

3.4.3 Growth of silicon crystals

During the production of pure silicon no special attention is paid to its crystallographic structure, and specific crystallisation techniques are usually required to obtain a material with appropriate semiconductor properties. The most common techniques produce silicon ingots with either a cylindrical or a square shape that need to be subsequently sliced into wafers. There are, nevertheless, techniques capable of directly producing thin sheets of crystalline silicon. Although ultra-pure polysilicon is the source material used by the microelectronic industry, it is relatively common among the PV industry (particular for square block techniques) to use discards from the former; examples of this are secondary grade polysilicon, the off-cuts of CZ ingots (tops and tails), off-spec ingots and wafers, etc. The price of discarded material is advantageous at present (approximately $15/kg) but its availability is subject to the internal demand of the microelectronic market. It is expected that an independent production of silicon feedstock for PV applications will be needed in the near future.

The most common crystallisation method used by both the microelectronic and photovoltaic industries is the *Czochralski (CZ) method* shown in Fig. 3.25, although the PV industry is increasingly shifting towards the directional solidification techniques described below. In the CZ crystal growth, silicon chunks are first melted at 1414 °C in a graphite crucible lined with high purity quartz. A small polysilicon crystal properly cooled is used as a seed to start the crystallisation process. The seed is carefully brought into contact with the melt and

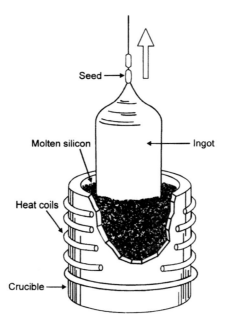

Fig. 3.25 Single-crystal silicon ingot growth by the Czochralski method

then pulled out very slowly. The temperature is tightly controlled so that silicon solidifies at the interface between the seed and the melt and the atoms arrange themselves according to the crystallographic structure of the seed. The crystal thus grows both vertically and laterally, aided by a rotation movement, yielding a cylindrical ingot of single-crystal silicon. The degree of purity can improve during the growth process since most impurities tend to segregate towards the liquid phase. Nevertheless, unwanted contamination can also occur during growth due to partial dissolution of the crucible that contains the molten silicon. A controlled amount of boron or phosphorus is usually added to the melt to dope the silicon p- or n-type. Rather than the elemental boron or phosphorus, accurately measured amounts of silicon heavily doped with those elements are added to the melt. The typical boron concentration used for solar cell applications is $1.5 \times 10^{16} \, cm^{-3}$, which results in a resistivity of $1 \, \Omega \, cm$.

The growth rate in the CZ method is about 5 cm/h and the cylindrical ingots are typically 1 m long, 20 cm in diameter (15 cm diameter has been common until recently and 30 cm diameter ingots are starting to be made) and 75 kg in weight. To increase the throughput, the crucible can be continuously replenished with molten silicon in some machines. A disadvantage of the CZ method is that the interaction between the molten silicon and the crucible introduces some contaminants, in particular carbon and oxygen (from the SiO_2 liner used in the crucible). The presence of oxygen in the crystal has been identified to be responsible for a degradation mechanism that has been observed in CZ silicon solar cells when they are first exposed to sunlight. Although this degradation is not very important (less than 1 efficiency point, stable after just a few hours) and it is not related to that happening in amorphous silicon, a number of research institutions have found ways of avoiding it. Crucible contamination can also be avoided by using magnetic fields to confine the melt, in a technique

Table 3.2 Conversion efficiency of solar cells made with different qualities of crystalline silicon, including Float Zone, Czochralski and multicrystalline silicon

CRYSTALLINE SILICON SOLAR CELLS (Illumination AM1.5G, 25 °C)				
Material	Cell structure	Organisation	Area (cm^2)	Efficiency (%)
FZ	Local P and B diffusions (PERL)	UNSW, Australia	4	24.7
CZ	Local BSF	ISE, Germany	4	22
CZ	screen printing, selective emitter	IMEC, Belgium	95	16.7
CZ	Laser grooved buried grid	BP Solar, Spain	143	16.7
CZ	industrial	several	100–140	13–15.5
mc-Si	phosphorus + boron diffusions	UNSW, Australia	1	19.8
mc-Si	textured surface	Sharp, Kyocera, Japan	100, 225	17.2
mc-Si	industrial	several	100–225	11–13

called magnetically-grown CZ; the resulting material has a very high quality, but it is not being produced commercially yet. Table 3.2 gives a summary of the state of the art for solar cells made from different kinds of silicon crystals. For CZ silicon the typical efficiency is 14% for cells made commercially by screen printing, and 15.5% for the more advanced buried contact technology. Higher efficiencies, up to 20% for 100 cm^2 cells and 22% efficiency for 4 cm^2 cells have been demonstrated in the laboratory.

The highest quality silicon crystals are obtained by using the *float-zone process*. In this method, the starting polysilicon is first given the shape of a cylindrical bar. The bar is then locally melted by a coil using radio frequency induction. By moving the coil, and hence the molten zone, along the bar starting from the seed end, the silicon adopts the crystalline structure. It is significant that the molten zone is self-supporting and is never in contact with a foreign material, thus avoiding contamination problems. A very high purity and also high structural quality can be achieved by performing several passes of the coil, thus maximising impurity segregation. The typical growth rate is 15–30 cm/h, and the typical ingot is 15 cm in diameter and 1 m in length. FZ silicon is the preferred material for the fabrication of high-efficiency cells, which have attained a 24.7% for a 4 cm^2 device.

The efforts of the photovoltaic industry to reduce costs and increase production rates have led to the development of new crystallisation techniques. The most successful ones produce large parallelepipeds of *multicrystalline silicon (mc-Si)*. It is possible to grow silicon ingots by simply melting the starting material, typically silicon scrap, into a crucible and carefully controlling the cooling rate (see Fig. 3.26). Alternatively, molten silicon can be poured into the square-shaped solidification crucible from a second container. Upon cooling, a directional solidification takes place and relatively large crystals grow in a columnar way. Note that a crystalline seed is not used and the nucleation of the silicon atoms commences in many places simultaneously, leading to a myriad of crystals (or grains) of arbitrary shape and crystallographic orientation. Because of its multiple-grained structure the material is called *multicrystalline*. Each grain is several millimeters to centimeters across, and internally it has the same structure as single crystalline silicon. The boundaries between the different grains are the most obvious imperfection in the material, but they are not the only ones. Microdefects are also common and contamination from the crucible can happen as well, not to mention the possible impurities present in the starting silicon. These crystallographic defects and impurities mean that mc-Si typically has a lower electronic quality than the material produced by the CZ method, leading to a typical loss of efficiency of 1% absolute and a wider spread in the production statistics; this difference is, nevertheless, narrowing rapidly. Mc-Si typically contains much less oxygen than CZ Si and does not show light-induced degradation. Several manufacturers use different directional solidification and casting methods to produce blocks of 45 × 45 cm^2 (up to 60 × 60 cm^2) cross-sectional area and 100–200 kg in weight. The typical crystallisation rate is 3.5 kg/h, and the growth cycle of a complete 160 kg ingot takes 46 h. This is, nevertheless, faster than the CZ method. The energy consumption is also more

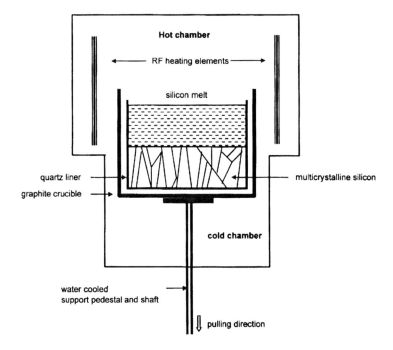

Fig. 3.26 Schematic diagram of a directional solidification furnace for the growth of multicrystalline silicon ingots

favourable for the directional solidification method: 10–15 kWh/kg, compared with about 100 kWh/kg for the CZ technique. Typical commercial mc-Si cells are about 13% efficient, with just over 17% having been demonstrated for large 100–225 cm^2 cells, and 19.8% for small 1 cm^2 laboratory devices.

The silicon ingots have to be sliced into wafers. Before this they are shaped to meet dimensional specifications. The cylindrical CZ ingots are usually reduced to a quasi-square shape. This implies a loss of about 25% of the material, but is necessary if a high packing factor of the cells in the module is required. The large cast silicon parallelepipeds are sawn into smaller bricks. In the case of mc-Si ingots, the shaping is also used to discard the peripheral regions that are usually heavily contaminated by the crucible, which represents approximately 15% of the ingot. Typical wafer sizes are 10×10 cm^2, $11.4 \times 11.4 = 130$ cm^2 or even $15 \times 15 = 225$ cm^2. The inner diameter saws used by the microelectronic industry have been replaced in the photovoltaic industry by multi-wire saw machines that can cut simultaneously whole blocks, thus increasing the throughput dramatically (Fig 3.27). An abrasive slurry helps the steel wires cut through silicon, a very hard material indeed. The cutting is very slow, with eight hours being typically needed to cut through a 10×10 cm^2 block. Despite this advanced technique, slicing remains as one of the most costly steps of the whole silicon solar cell fabrication. Even if very thin wires are used approximately 30% of the silicon is wasted as saw dust, or *kerf loss*.

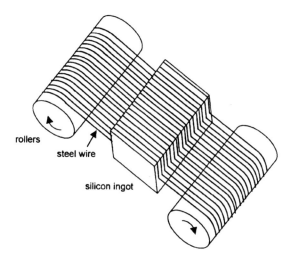

rollers

steel wire

silicon ingot

Fig. 3.27 Multi-wire saw

It is interesting to evaluate the yield that can be expected from each kg of silicon in ingot form considering that the wafer thickness is typically 300 μm. We can use the value of silicon density ($2.328 \, \text{g/cm}^3$) to calculate the weight of the wafer per unit area ($0.07 \, \text{g/cm}^2$ – that is, a $100 \, \text{cm}^2$ wafer weighs 7 g). Considering a typical kerf loss of 200 μm the resulting yield comes out to be approximately $0.8 \, \text{m}^2$ per kg. The impact of silicon material on the final cost of the solar modules can be estimated by using a typical conversion efficiency in the range 10–15% and the standard solar radiation of $1 \, \text{kW/m}^2$. With 1 kg of crystallised silicon, it is possible to produce 80–120 W_p. If the contribution of the silicon wafers to the module cost is to be kept at, for example, $1/W_p$ (current module costs are in the range of $4/W_p$), the cost of the silicon crystal must be well below $80–120/kg since allowance must be made for slicing. There is a trend towards cutting thinner wafers – 150 μm thick, for example – which would increase the wafer yield per kg. Nevertheless, advanced device fabrication and wafer handling still need to be developed.

Considering how cumbersome the sawing of ingots is we should not wonder at the variety of methods that have been investigated to grow the silicon crystal directly in a sheet form. Several techniques are still being intensively developed. A good example is the EFG method (Edge-defined Film-fed Growth), shown in Fig. 3.28. The molten silicon ascends by capillary action through a thin slot in a graphite die. The crystallisation is initiated with a seed that is pulled upwards, thus growing a ribbon several meters long (up to 4.6 m) and 300 μm thick. In commercial production a tubular octagon is grown using the appropriate die shape, and eight ribbons, each 10 cm wide, are subsequently cut using a laser. An interesting feature of most sheet-growth techniques is that they can have a high production rate – for example, $160 \, \text{cm}^2/\text{min}$ for the EFG method. Experimental cells made with EFG silicon have reached 16%, with average

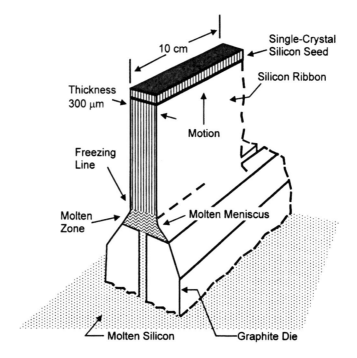

Fig. 3.28 Ribbon growth by the EFG method

commercial cells being 11–13% efficient. Other sheet growth methods have also demonstrated similar efficiencies. Wafers grown by the ribbon growth technologies have so far had a relatively small impact on the market (1–2%), although their share is likely to increase in the near future.

There are other approaches to circumvent the problems associated with the crystallisation and sawing steps. One of them is based on the epitaxial growth of silicon layers from a liquid phase. This method starts by alloying silicon chunks with a metal like tin or indium at a temperature of about 900 °C (note that the melting point of the alloy is significantly lower than that of pure silicon). Controlled cooling makes the silicon precipitate from the liquid phase on top of a substrate, which can be a silicon wafer (from which the grown layer is subsequently detached) or a foreign substrate such as a low cost ceramic. The thickness of the epitaxial silicon layer ranges between 25–250 μm, the aim being to obtain high efficiencies in the thinner layers, thus reducing the use of silicon material by approximately 20 times compared to wafer technologies. Commercial size silicon film cells on foreign substrates have already demonstrated over 10% efficiencies. We shall discuss this technique further in section 3.5. Another approach, which is being investigated intensively, aims at depositing truly thin films (2–20 μm) of silicon, which commonly have a polycrystalline crystallographic structure (as distinct from both multicrystalline silicon, which has much larger grain sizes, and amorphous silicon, which does not show a well-defined crystallographic arrangement at a macroscopic level). Some thin-film polycrystalline silicon techniques have already reached the 10% efficiency mark for small laboratory devices. It is reasonable to expect that these new technologies will have an impact on the commercial production of solar cells in the near future. Nevertheless, the technologies based on ingot silicon are very well established and likely to remain dominant for several years. The recent expansion of production capacity experienced by the PV industry is mainly based on these conventional technologies. One half of the total wafer production in 1998 for solar cells was multicrystalline silicon (including ribbon) technologies, with the rest being monocrystalline silicon produced by the Czochralski method.

3.4.4 Typical solar cell fabrication process

To transform a silicon wafer into a solar cell, the wafer is subjected to several chemical, thermal and deposition treatments. The cross section of a silicon cell (Fig. 3.29) shows the different layers that need to be formed, and points out the two most essential layers: an n-type layer to form the p–n junction and two metal layers to form the electrical contacts. In a typical industrial fabrication sequence there are a number of additional steps, as follows:

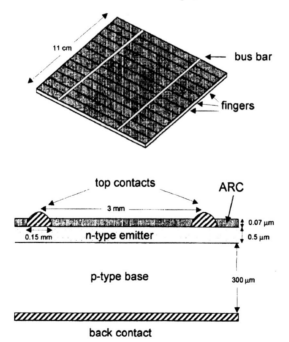

Fig. 3.29 The structure of a silicon solar cell (not to scale). Typical dimensions are given for the metal grid (2 mm wide bus bars, 150 μm wide fingers), and the thicknesses of the antireflection coating (70 nm) and the diffused region (0.5 μm)

1. surface cleaning and etching (may include texturing);
2. phosphorus diffusion for p-n junction formation;
3. front and back metal contacts;
4. antireflection layer deposition.

Immediately after cutting, the silicon wafers are dirty and their surface is heavily damaged. Several wet cleaning steps and chemical etches are used to remove a few microns of silicon and leave the wafers ready for the phosphorus diffusion. Although silicon is an extremely hard and chemically resistant material, it can be etched easily in hot concentrated solutions of sodium or potassium hydroxides. In the case of monocrystalline silicon wafers that have an appropriate crystallographic orientation defined as (100) plane, it is possible to use a modified KOH solution that etches silicon preferentially in a particular crystallographic direction and exposes the planes of the crystal that have more closely packed atoms (the (111) planes). The result (Fig. 3.30) is a *textured* surface completely covered by many small pyramids of random size (the base is typically in the range of 3–15 μm that help to reduce the reflection of the silicon wafer. The angle formed between the (111) and the (100) planes is 54.74°. Which results in the height of the pyramid being 0.707 times the side of the base and the area of the four triangular faces being 1.73 times the area of the base. Although the increased surface area of textured surfaces can be detrimental to the voltage of the solar cell, the overall effect is positive, due to a marked increase of the photogenerated current. Textured surfaces have a black appearance when they are wet or behind a glass indicating that little of the incident light is reflected). When light rays fall on the facets of the pyramids a fraction of them bounce off towards neighbouring pyramids, giving it a second chance to penetrate into the solar cell. The reflectivity of bare silicon is thus decreased from $R = 0.35$ for a polished surface to approximately $0.35 \times 0.35 = 0.12$ for a textured one. When the textured wafers are encapsulated under glass the reflection loss can decrease to 2% ($R = 0.2$ for a glass-encapsulted silicon surface), plus a 3.5% reflection off the top of the glass plate. Alternative surface texturing schemes are being developed for multicrystalline silicon; some of them are based on the mechanical action of shaped blades or wheels, while others use reactive ion etching, a plasma etching technique.

The starting silicon wafers are usually p-type, that is, boron-doped. It is then customary to form the p-n junction by introducing phosphorus, a n-type impurity, from the front surface. At sufficiently high temperatures (870 °C is

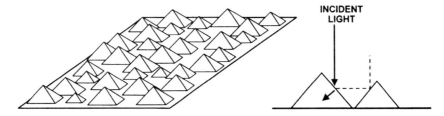

Fig. 3.30 Pyramidally textured silicon surface

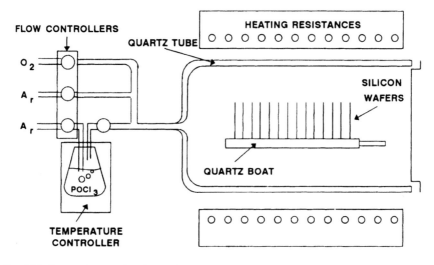

Fig. 3.31 Open tube quartz diffusion furnace

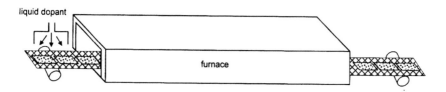

Fig. 3.32 Conveyor belt furnace for phosphorus diffusion

typical of solar cell processing), phosphorus atoms can diffuse into the solid silicon wafer. For a typical diffusion time of 15–30 min, the penetration depth is very small (approximately $0.5\,\mu m$), as required for optimal solar cell operation. The conventional way of performing a phosphorus diffusion is to use a quartz diffusion furnace, as shown in Fig. 3.31. A common dopant source is a liquid chemical containing phosphorus ($POCl_3$), which is conveniently carried into the furnace by bubbling nitrogen through it. In addition, oxygen is injected into the furnace so that it reacts with the $POCl_3$ and forms phosphorus oxide (P_2O_5). At the surface of the wafers the P_2O_5 turns into silicon dioxide (SiO_2) and atomic phosphorus, which can thus diffuse into the wafer. The oxide that is left on the wafers is usually removed chemically after the diffusion. Although the open-tube quartz furnaces are used by some companies, the vast majority of solar cell manufacturers use a conveyor belt furnace to perform the phosphorus diffusion step in a continuous, high throughput way (Fig. 3.32). Typically the wafers are sprayed with a liquid that contains phosphorus immediately before being fed into the furnace. Alternative dopant deposition techniques include screen printing, spinning or chemical vapour deposition of phosphorus oxide. A typical industrial furnace is 0.6 m wide, 10 m long (6 m hot zone) and can process 1200 wafers/h.

The diffusion of phosphorus usually extends around the edges of the wafer and, in some cases, to the rear surface. To remove this unwanted n-type region, the wafers are coin stacked and placed in a plasma etching machine. An additional fabrication step that is known to improve cell performance is the formation of a p^+ region at the rear surface. As we have seen in section 3.3, this feature creates a back surface field that decreases the chances of carriers recombining at the back surface. It is straightforward to create such a p^+ region by depositing an aluminium layer and alloying it with the silicon at a temperature of about 800 °C; the aluminium is a p-type dopant, just like boron, and it creates a p^+ region. Besides reducing surface recombination, the aluminium alloying step can help to remove from the wafer damaging impurities like iron, chromium, etc; this is called *gettering* and it is particularly important for multicrystalline silicon. A significant gettering action also takes place during the phosphorus diffusion step. The aluminium alloying technique is well know and simple (it can be deposited by screen printing and alloyed in conveyor belt furnaces), and it is used already by some manufactures. Its importance is growing due to the trend to reduce the thickness of the wafers.

Electrical contacts are usually formed by screen printing. This technology is inexpensive, simple, and can be automated. The screen consists of a mesh of wires embedded in an emulsion, as shown in Fig. 3.33. This emulsion is photographically patterned and removed from the places where metal is to be deposited. A paste containing the metal is squeezed through the screen onto the wafer, thus depositing the metal grid. Silver is the metal typically used for contacting the phosphorus-diffused region at the front-surface, while a mixture of aluminium and silver is used to make contact to the bulk p-type silicon wafer at the rear. Upon firing the organic solvents evaporate and the metal powder becomes a conducting path for the electrical current. The paste also contains small amounts of glass that provide a good adhesion to the silicon surface. The firing is done in conveyor belt furnaces similar to that shown in Fig. 3.32 at a temperature of about 700 °C for a few minutes. The typical width of the screen printed metal grid fingers lies in the range 150–200 μm and their separation is the order of 3 mm. Two 2 mm wide metal stripes called bus bars complete the grid and bring the total shading loss to 9%, approximately. High-efficiency and concentrator cells need much narrower lines to minimise shadowing losses. The 10–40 μm line widths frequently encountered in their metal grids are made by photolithography, a technique characteristic of microelectronic fabrication. The metals are then deposited in a high-vacuum machine either by electrical resistance heating, electron-beam evaporation, or by sputtering. Small area (4 cm^2) laboratory cells can have a shading loss of 5% or less.

A thin layer of a transparent material which acts as *antireflection coating* is usually deposited after the formation of the metal contacts (or sometimes even before the metal deposition). This dielectric material should have an optimum value of refractive index intermediate between those of silicon and glass,

$$n = \sqrt{n_0 n_{Si}} = 2.4$$

where $n_0 = 1.5$ for glass and $n_{Si} = 3.7$ for silicon. The optimum thickness d is obtained from the condition that produces a minimum reflectivity at the wavelength within the useable part of the solar spectrum (350–1100 nm) where the photon flux is greatest, $\lambda = 650$ nm, approximately

$$d = \lambda/n_4 = 70\,\text{nm}$$

where $n = 2.4$. An appropriate material that is commonly used by the PV industry is titanium oxide (TiO_2). Tantalum pentoxide and zinc sulphide are also suitable for optimised antireflection coatings. Another material that is particularly interesting is silicon nitride (Si_3N_4) because it has been shown to have beneficial effects on the electronic properties of the silicon wafers. A single-layer antireflection coating can decrease the reflectivity of a flat silicon

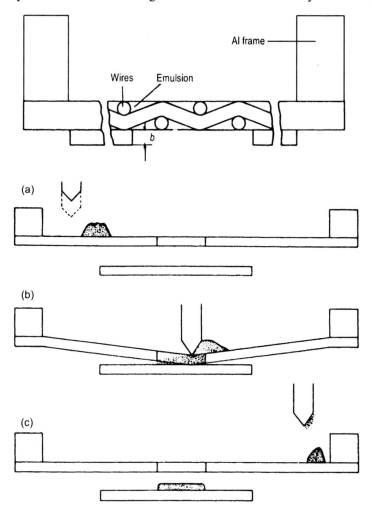

Fig. 3.33 Screen printing technique

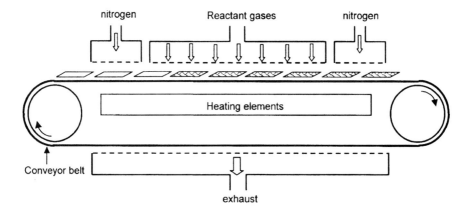

Fig. 3.34 Atmospheric Pressure Chemical Vapour Deposition (APCVD) system for the deposition of antireflection coatings, for example titanium dioxide

surface, weighted over the useable part of the solar spectrum, to about 9%, including 3.5% from the glass. Total reflection plus metal grid shading losses for encapsulated multicrystalline silicon cells are approximately 18%. Mono-crystalline silicon cells are usually pyramidally textured and in this case the total reflection plus shading loss are lower, approximately 14%. Several methods can be used for the deposition of antireflection coatings, including spraying, screen printing, spinning, etc; the most commonly used by the PV industry is the chemical vapour deposition of organic titanium oxide compounds. As Fig. 3.33 shows, this is done in conveyor belt reactors at atmospheric pressure (APCVD) and temperatures in the range 300–400 °C.

Although it is broadly used, screen printing is not the only technique available for metallisation. The deposition of nickel from a chemical solution has been used in the past and is at the core of a new technology that has produced the highest solar cell efficiencies in an industrial production, the Laser Grooved Buried Grid (or buried Contact) technology. After a light phosphorus diffusion and dielectric deposition (silicon nitride), a high-power laser is scanned through the surface of the wafer to create shallow grooves following the pattern of the metal grid (the grooves are typically 30–50 µm wide, with a similar depth). The grooves then receive a second phosphorus diffusion and are subsequently filled in with nickel, copper and silver. The fabrication sequence is more complex than for screen printed cells, but it resolves quite brilliantly the multiple trade-offs between metal contact resistance, phosphorus doping of the n-type layer and surface passivation. The result is enhanced conversion efficiencies, which are in the 15–16% range for production cells, with top efficiencies up to 16.7%.

A summary of the conversion efficiency achieved by several different cell designs is given in Table 3.2. It can be seen that both the buried grid technology and advanced screen printing techniques are capable of efficiencies close to 17%, although the latter gives significantly lower efficiencies in large production scale.

3.4.5 Module fabrication

After testing under standard conditions (see Sec. 3.3.3) and sorting them in different classes to match their current and voltage, about 36 cells are typically interconnected in series and encapsulated to form a module. Different module sizes are manufactured for different applications; the largest are for grid connected power plants and measure $2\,m^2$. The encapsulating materials must satisfy a number of very strict requirements, since the modules should last 20 years or more. A reliable module construction, such as that in Fig 4.3, has the following components.

1. Front cover: tempered glass. Although low iron glass has been used in the past, recently cerium oxide is being incorporated into the glass to reduce the transmission of ultraviolet light (of wavelength less than 360 nm) and reduce the possible degradation of the encapsulating materials.
2. Encapsulant: a transparent, electrically insulating, thermoplastic polymer. The most widely used is EVA (ethylene vinyl acetate). The typical thickness is 0.46 mm.
3. The solar cells and the metal interconnects.
4. Back cover, usually a foil of Tedlar, Tefzel or Mylar (polyester). A very thin aluminium foil can be included in the sandwich to provide a barrier against humidity.

All these layers are laminated by applying heat (at approximately 150 °C) and pressure under vacuum. Primers are used to improve adhesion between the layers. The edges are sealed with a neoprene gasket and, in some cases, protected with an aluminium frame. An alternative encapsulation scheme uses a second glass plate as back cover and a sealing gasket without a metal frame, since the latter is rendered unnecessary by the extra mechanical strength of the two laminated glass plates. A specialised encapsulation scheme is needed for bifacially sensitive modules, which have a transparent back cover (for example, transparent tedlar) and can collect up to 30–50% extra light from the albedo (background reflection). Once finished, the modules are tested and rated according to their power output. It should be noted that the cell interconnection produces a moderate loss of efficiency, for example a module efficiency of 14% typically results from using 15% cells; bigger losses can occur if the cell fabrication produces a broad spread of cell efficiencies.

Summary

Crystalline silicon solar cell technology has been described in steps from sand to final module fabrication. Much of this technology has been adapted from the microelectronics industry but specific methods to produce multicrystalline silicon ingots, ribbons and sheets and to fabricate solar cells in large scale have been developed. Further research is still needed to replace the microelectronic grade feedstock with a lower cost solar-grade silicon.

Cell and module-fabrication technology is well established and reliable. Advanced solar cell device designs and fabrication processes are being developed, an the laser grooved buried contact technology is a good example of successful transfer from the research laboratories to industrial production. Efforts to obtain higher conversion efficiencies are also taking place in the area of screen printing technology. Multicrystalline silicon materials and cells are improving steadily and gaining a greater share of the market.

3.5 THIN-FILM SOLAR CELLS

3.5.1 Introduction

In comparison with the crystalline silicon cells that were discussed in section 3.4, thin-film technology holds the promise of reducing the module costs through lower material and energy requirements of the manufacturing process. In addition, integrally connected modules are produced directly without the costly individual cell handling and interconnections.

Amongst the large number of possible thin-film cells, a few stand out as candidates for commercial production in the near future. The criteria for commercial viability which we shall now discuss are based mainly on economic considerations and are obviously generalisations. Nevertheless, they suggest technical goals whose achievement would open up very large markets for thin-film solar cells.

Solar radiation has a low power density, and hence the area of PV modules needed to produce significant power output can become quite large. Many costs associated with a photovoltaic generating system increase in proportion to the area of the system. Most module production costs are area-related, and the PV system costs of land, array structures, wiring and transport are all area-related and can dominate the total cost of the system if the efficiency of the modules is low. It is generally considered that module efficiencies over 10% are needed if the system cost is to be held to acceptable levels.

Photovoltaic systems have large initial capital costs but small recurrent costs for operation and maintenance. The cost of delivered energy thus varies inversely as the lifetime of the system. The present PV modules based on silicon wafers exhibit lifetimes of 20–30 years and the target for thin-film cells must be similar.

The annual rate of production of PV modules will reach a few million square metres by the end of the century, and probably a few hundred million square metres in the next 40–50 years. Cells which contain scarce materials could thus be subject to severe constraints due to the availability of these materials at a reasonable price.

The move towards the use of thin-film solar cells is driven by the need to reduce module costs. The cells must therefore be capable of being manufactured in large volumes at low cost. To avoid costly handling, the production process should result in an integrally interconnected module, rather than individual cells requiring separate interconnections.

Table 3.3 Best results for thin-film solar cells (efficiencies in % under standard AM 1.5 illumination)

Material	Commercial products	Best large area ($\sim 0.1\,m^2$)	Best R&D small area	Theoretical limit
Amorphous silicon	5–8	10	13	$\sim$20
Thin-film silicon	11	12	16	$\sim$25
Copper indium diselenide	8	14	16	$\sim$21
Cadmium telluride	7	11	16	$\sim$28

The final requirement for the solar cell is that it results in no environmental hazard. The hazards must be assessed from the initial mineral extraction, through refining, cell manufacture, PV system operation and final disposal of the modules. The hazards associated with present PV materials are discussed in Chapter 6.

Four types of thin-film cell have emerged to be of likely commercial importance in the next few years (Table 3.3). These are the amorphous silicon cell, in a multiple-junction structure, thin multicrystalline silicon films on a low-cost substrate, the copper indium diselenide/cadmium sulphide heterojunction cell (or variants of it), and the cadmium telluride/cadmium sulphide heterojunction cell.

All of these cells have active layers in the thickness range 1000–10 000 nm, and all are made by processes which are capable of large-volume, low-cost production. The materials and the processes were chosen because they had this capability, and the major R&D effort has been to increase the efficiency of the cells and the reproducibility of the product.

Other approaches to thin-film cell manufacture are also being pursued and should be mentioned here. For example, polymers can be doped both p-and n-type, and polymer diodes have been used as luminescence emitters. These materials have been studied for use as solar cells and quite encouraging results have been obtained. They can be very cheap, but need to show efficiencies of at least 5% or so before one can envisage any commercial use.

3.5.2 Amorphous silicon cells

Amorphous silicon (a-Si) differs from crystalline silicon in that the silicon atoms are not located at very precise distances from each other and the angles between the Si–Si bonds do not have a unique value. The interatomic distances and the bond angles have a range of values with any particular Si–Si distance or bond angle being randomly located within that range.

This randomness in the atomic arrangement has a powerful impact on the electronic properties of the material. It becomes a direct-gap material with an optical energy gap of about 1.75 eV, and its band structure shows a very high density of states within the energy gap, caused largely by incomplete bonding. It was found in 1969 that the incorporation of hydrogen in amorphous silicon

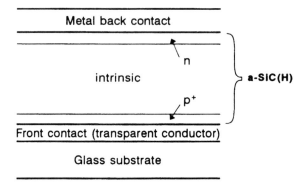

Fig. 3.35 The structure of amorphous silicon p-i-n solar cell

could passivate the incomplete bonds and reduce the density of states in the bandgap to such an extent that it became possible to make the material n-type or p-type by the incorporation of phosphorus or boron. Once it became possible to make p-n junctions in a-Si(H), its high absorption coefficient and ease of manufacture made it an attractive material for solar cells. Both p and n materials have poor transport properties, and simple p-n junction cells have low efficiencies. As we have seen in section 3.4.1, the incorporation of intrinsic a-Si(H) improves the transport properties, and the development of p-i-n junction cells led to rapid improvements in performance. The i-layer is not in fact completely intrinsic, but is slightly n-type. It is therefore preferable to have light enter the cell through a very thin p-type layer so that the region of maximum photogeneration in the i-layer is also the region of highest electric field, between the heavily doped p^+-layer and the very lightly n-doped 'i'-layer.

The optimum structure of the a-Si(H) cell is the p^+-i-n, which was already discussed in section 3.3, has a transparent conducting contact to the p^+-layer and an ohmic contact to the n-layer, as shown in Fig. 3.35. Unfortunately, an optically transparent highly conductive p-type material has not yet been discovered. The transparent conducting material used is usually tin oxide, which is n-type. However, the highly doped TCO (transparent conducting oxide) layers form a tunnel junction with the p^+ a-Si(H) layer, and the resistance to current flow is very small.

Having determined the optimum structure, the manufacturing process sequence can be defined (Fig. 3.36). Starting with glass as a low-cost transparent, weatherproof substrate, a layer of highly conductive optically transparent tin oxide is deposited, then a highly doped p-layer of a-Si sufficiently thin to absorb little light, then the undoped, but slightly n-type intrinsic layer of a-Si, then a thin conductive n-layer of a-Si and finally a metallic contact layer. This contact layer should form an ohmic contact to the n-layer of a-Si and ideally should also be highly reflective so that any light not absorbed in passing through the cell is reflected back for a second pass through the i-layer. Silver would be the best material except for its cost and aluminium or aluminium

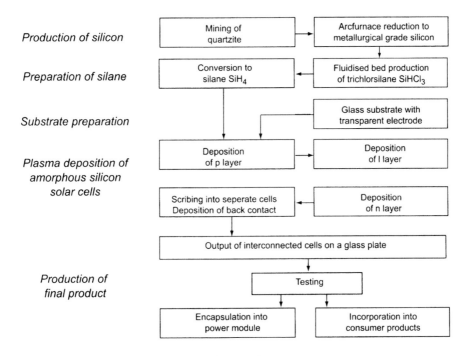

Fig. 3.36 Production of amorphous silicon PV modules

alloys are usually used. The original production process which produced the a-Si/H$_2$ alloy, needed for semiconductor-grade amorphous silicon, was the plasma decomposition of silane (SiH^4). This is still the best process, producing the most efficient cells, although other processes such as sputtering have been studied in some detail, and elements such as fluorine have also been proposed as passivators of the dangling bonds.

As the quality of the a-Si material improved through the 1970s, it became apparent that the action of light absorption by the i-layer created additional defects, increasing the density of trapping and scattering states and reducing the efficiency of the cells. This effect—named the *Staebler-Wronski effect* after its discoverers—is dependent on the total number of photons absorbed. It therefore depends on the intensity of the light to which the cell is exposed, the duration of the exposure and on the thickness of the i-layer. Exposure to room lighting, as with solar calculators, etc., has only a small effect, but bright sunlight reduces the efficiency considerably over timescales of months. This instability has serious consequences for the commercial viability of a-Si as a power producer and it now seems unlikely that the single-junction a-Si cell will be used in such applications.

Since the Staebler—Wronski effect depends on the thickness of the i-layer, it can be alleviated by using multiple-junction structures, in which the absorption of light is split equally between two or three separate i-layers. The simplest of such structures is just a p-i-n/p-i-n stack of a-Si(H), shown in Fig. 3.37, in which

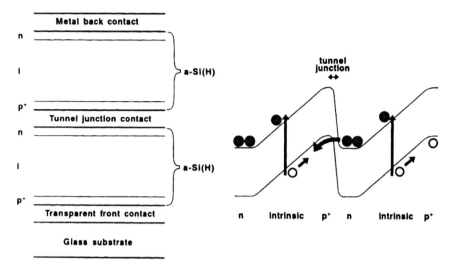

Fig. 3.37 The structure and band diagram of the amorphous silicon p-i-n/p-i-n solar cell

the overall thickness of the cell is maintained at about 1000 nm with the front i-layer about 300 nm and the back i-layer about 700 nm to produce equal currents from each half. Cells with these structures have been produced commercially and found to be significantly more stable than the single-junction device. Manufacturers claim that the stability is sufficient to allow these cells to be commercially viable as power generators.

Amorphous silicon forms good alloys with both carbon and germanium. The bandgap of the alloy a-SiC(H) is increased in proportion to the fraction of carbon whilst that of a-SiGe(H) is reduced in proportion to the fraction of germanium. Both alloys can be produced by the same plasma deposition process which deposits a-Si(H) with just the feed silane being replaced by a mixture of silane/methane (SiH^4/CH^4) or silane/germane (SiH_4/GeH_4). This gives the possibility of many multi-junction structures, for instance a p-i-n front cell of a-SiC(H), a middle p-i-n cell of a-Si(H) and a back p-i-n cell of a-SiGe(H) shown in Fig. 3.38. Such a cell has been produced with an efficiency of 13.3%, compared to the best single-junction efficiency of 12%.

The principal technical difficulty with these cells is the formation of the tunnel junction between the n-layer of the upper cell and the p-layer of the lower cell. Good results can be achieved in the laboratory on small-area cells, but it is a major challenge to achieve similar results reliably and repeatably in commercial production on large-area modules.

3.5.3 Thin polycrystalline silicon on low-cost substrates

There is good theoretical justification for using thin base layers in silicon solar cells because the dark saturation current I_o decreases with decreasing base-layer

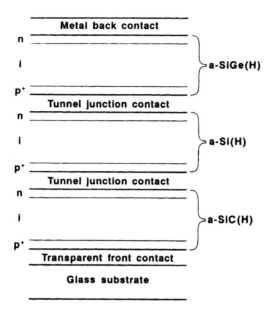

Fig. 3.38 Triple-stack amorphous silicon solar cell

thickness, leading to higher open-circuit voltage values for thin cells. The problem, at first sight, is that complete optical absorption requires thicknesses of a few hundred micrometres. As we have seen in section 3.3, it is nevertheless possible to maintain optical path lengths of hundreds of micrometres with silicon layers of tens of micrometres thick by using light-trapping techniques.

Light incident on a flat planar cell will enter through an antireflection (AR) coating and an exponentially decreasing proportion will travel through the cell to the back contact. Some light incident on the back contact will be reflected back through the cell for a second pass in which it will be either absorbed or leave the cell via the AR coating. If the top and back surfaces are textured, rather than being flat, the light entering through the AR coating will be refracted off-normal, and light reaching the back will be scattered, as will light reaching the top surface after reflection from the back. In this way, the optical path length of a typical light ray normally incident on the cell can be up to 20 times the cell thickness. Complete optical absorption can then be assured even for cell thicknesses of only a few tens of micrometres.

There are other benefits to such thin silicon cells. The diffusion length need be only 50–80 µm, so lower quality material can be used, and higher doping levels can be tolerated, giving higher open-circuit voltage. Because the optical absorption, and hence photogeneration, is taking place within or close to the junction depletion region, carrier collection efficiencies are high, giving high short-circuit currents. The reduced constraints on material quality mean that film deposition of silicon layers on low-cost substrates can be considered without sacrificing much in the performance of the cells. This is therefore a very promising route for the production of efficient cells at low cost.

There are a number of key areas to the successful implementation of the concept. One needs a low-cost substrate which has a good thermal expansion match to silicon, to avoid strain in the silicon film after processing, and one needs a silicon film-deposition process which gives high yields of suitable materials and potentially high throughput at acceptable capital cost.

The growth of silicon films from saturated solutions of silicon in molten tin by *liquid phase epitaxy* (LPE), already mentioned in section 3.4, has been found to meet the requirements for a commercial production process. The films as-grown have a columnar polycrystalline structure and post-processing is needed to increase polycrystallite size and passivate the grain boundaries. After this treatment, the p-type silicon film is processed by conventional diffusion techniques to give an n-p junction, then AR coated and top contacted in the usual way. The light trapping arises from the topology of the substrate surface, and the optical properties of the back contact metallisation. Efficiencies up to 16% have been observed in small-area cells and about 12% in commercial 10 cm × 10 cm cells. Other deposition processes have been developed for other designs of silicon film, and are being actively investigated for commercial production.

The deposition on to small areas cannot provide a commercial process as it effectively produces the equivalent of a wafer silicon cell, which must be interconnected at the subsequent module production stage. It is possible to envisage a module-sized ceramic substrate with cells deposited simultaneously on the substrate with a subsequent metallisation/interconnection step for the whole module. The key to the realisation of this advance is the availability of ceramic substrates of suitable size, and the ability to achieve uniform deposition over the whole surface. There are reports that module scale deposition has been achieved in the research laboratory. If these should become commercially available, then thin silicon technology could be an important route to the target module price of $1 per peak watt.

3.5.4 Copper indium diselenide cells

Copper indium diselenide (CIS) is a semiconducting material which can be either n-or p-type and has a direct optical absorption with the highest absorption coefficient measured to date. The electronic properties of CIS depend critically on the copper/indium ratio and good control of the stoichiometry is essential for efficient devices.

It is possible to make p-n homojunctions of CIS, but these are neither stable nor efficient, and the best devices to date are heterojunctions with cadmium sulphide (CdS). There is quite a good match between the lattice constants and electron affinities of CIS and CdS so the recombination at their interface is not excessive. CdS can only be grown as n-type material, so the CIS must be p-type. It is fairly easy to make ohmic contact to CdS, particularly when it is heavily doped, but contacts to the p-type CIS are more difficult. Gold makes a good ohmic contact but is too expensive for commercial use. Molybdenum is the metal most commonly used, and makes reliable contacts to heavily doped p-type CIS. For efficient devices, the CIS needs to be fairly lightly doped near the junction, so that the depletion region extends throughout the volume where photogeneration takes place. This ensures field-aided collection of the photo-generated carriers and hence high short-circuit currents.

CIS has a bandgap of 1 eV and grows as a columnar polycrystalline film. The CdS also grows, to some extent topotaxially, as a columnar polycrystalline film on top of the CIS. The open-circuit voltage of the cell is thus very dependent on the recombination currents near the junction, due to the grain boundaries and the interface. Good quality CIS has grain sizes of a few micrometres and diffusion lengths of around 2 μm, so passivation of the grain boundaries is essential to reduce the recombination currents. An oxidation treatment after deposition, for instance heating in air, can give significant improvements in performance, due to grain boundary passivation, and clearly a deposition method which gives large grains with minimum grain boundary states is advantageous.

CdS has a bandgap of 2.4 eV so it will absorb strongly all incident radiation in the green/blue end of the spectrum. The optimum devices use a very thin (30 nm) layer of CdS with a wide-gap window layer of highly conductive material. Zinc oxide has been found to be a very suitable material for the window layer, so the structure of the CIS cell is that shown in Fig. 3.39.

The production sequence of CIS modules is shown in Fig. 3.40. In the first ten years of R&D on CIS cells, the deposition method used by all workers was vacuum evaporation of the elements. It was quickly discovered that mainten-ance of stoichiometry depended on providing the correct copper/indium ratio and supplying an excess of selenium vapour pressure above the growing film surface. A later development with better potential for commercial production was to deposit metallic indium and copper on to the molybdenum contact metal and then selenise the indium/copper layers by passing hydrogen selenide gas over the surface at about 400 °C. This selenisation process gives high-quality

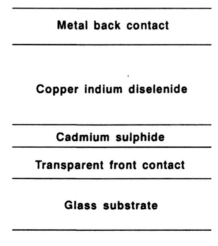

Metal back contact

Copper indium diselenide

Cadmium sulphide

Transparent front contact

Glass substrate

Fig. 3.39 Structure of CIS solar cell

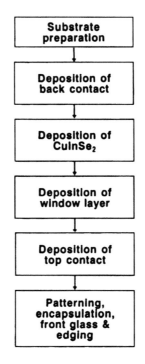

Fig. 3.40 Production of CIS modules

material and can produce uniform films over large areas, suitable for module production. The incorporation of selenium into the indium/copper layer causes a large increase in volume, and can result in large stresses in the CIS film, even to the extent of delamination of the film from the substrate. These problems have been addressed by careful control of the details of substrate preparation, molybdenum deposition, sputtering of the copper and indium, and control of the selenisation process. The addition of some gallium at the Mo/CuIn interface is also found to improve the adhesion quite markedly.

The incorporation of gallium into CIS has a number of benefits, apart from improved adhesion of the film to the molybdenum. Gallium substitutes for indium to give a true alloy—copper indium/gallium diselenide (CIGS)—with a bandgap which increases with increasing gallium fraction. As the bandgap increases, the open-circuit voltage increases, giving better fill factors and fewer cells per module. For CIS, the voltage at maximum power is usually only about 300–350 mV, which is inconveniently low. CIGS with 10–20% gallium can raise this to values of 450–500 mV, about the same as crystalline silicon cells, and, with fewer cells to be integrally interconnected in module production, the processing is more reliable and less costly. CIGS with a bandgap of greater than 1.1 eV also avoids problems with the free-carrier absorption in the zinc oxide window. The conductance of zinc oxide is due to free electrons and the long-wavelength absorption due to these free carriers begins at about

1.1 μm (1.1 eV). It is possible to reduce the free carrier absorption by improving the quality of the ZnO film, but it cannot be eliminated. However, with CIGS, the problem no longer exists since photons of energy of 1.1 eV or less are not absorbed by the CIGS, and the current lost by the increased bandgap is more than compensated for by the increased voltage and fill factor.

A considerable effort has been devoted to scaling up the CIS technology to module size, mainly by Siemens Solar Industries (formerly ARCO Solar). Modules have been in precommercial production for some time, and subjected to rigorous indoor and outdoor testing. A range of module sizes is now available commercially with efficiencies around 8%, which appear to be stable over periods of some years. Other manufacturers are now in or close to commercial production, and one can expect to see a number of products competing strongly within the thin-film market. As the uniformity of the deposition improves with learning and 'tweaking' over time, the efficiency of the modules will increase to 12–14%. Provided that the modules have demonstrated good stability by that time, the CIGS technology could begin to challenge the wafer silicon technology as the standard for the whole range of applications.

3.5.5 Cadmium telluride cells

It has been known since the 1950s that cadmium telluride has the ideal bandgap for a solar absorber material and the research on these cells can be traced back to that time. These efforts, however, were small and uncoordinated and severely hampered by the low level of materials technology for the II/VI materials. As the II/VI materials technologies improved over the years, it became clear that cadmium telluride performance was limited by a high level of defect states close to the middle of the bandgap, forming very efficient recombination sites. The achievement of the late 1970s was to develop post-deposition treatments which greatly reduced the density of these recombination states, and the achievement of the 1980s has been to capitalise on this earlier work to develop efficient cells and efficient stable modules.

There are a number of low-cost techniques which can be used to deposit cadmium telluride and they can all, after a post-deposition treatment, yield high-quality material and efficient cells. The choice of technology for commercial production can then be made on the basis of its fitness for the expected markets, environmental considerations and the availability of skills, capital, etc., within the production company.

There are at present three companies close to production of CdTe devices, Matsushita in Japan, BP Solarex in the USA and Antec in Europe. Matsushita have developed screen printing technologies for all of the films in the cell, back and front contacts, cadmium telluride and cadmium sulphide. Screen printing of solar cells offers a technology with low-cost production at quite low production rates. The rate-limiting step is the heat treatment of the printed inks after deposition, which requires fairly high temperatures (around 500 °C) for periods of an hour or more. The key to the success of the screen printing process

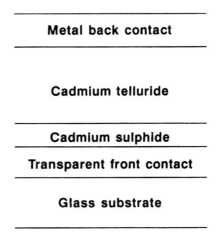

Fig. 3.41 Structure of CdTe solar cell

seems to be the use of elemental cadmium and tellurium in the inks, rather than CdTe powder. The Cd and Te react during the sintering process to produce CdTe and the heat of reaction is high enough to promote the growth of high-quality layers. These screen-printed cells have been sold for some years for use in solar calculators, but there are problems in scaling up to power modules because of the large degree of contact obscuration (40%) in this technology, although modules with an aperture efficiency of 6% have been produced and tested in outdoor conditions for some years, and are commercially available in limited quantities.

BP Solarex has developed the electroplating technology originally devised by Monosolar, a company which they purchased in 1984. Electroplating is a particularly favourable technology for CdTe cells because of its very high utilisation of material and low capital cost of the equipment relative to vacuum-deposition processes. It is relatively easy to scale up the process to produce module-sized layers and it should be capable of low-cost production at low volumes. BP Solarex have produced modules of area about $0.1\,\text{m}^2$ with efficiencies over 10% and 1 cm² cells with efficiencies of around 13%. Outdoor stability tests have been performed in Spain and application development tests elsewhere with excellent results.

There is a variety of vacuum deposition techniques which have produced good results in the research laboratory on small-area cells, but only one of these—closed space sublimation—is being considered for potential commercial development. It is envisaged that ultimately thin-film solar cells will be produced in a facility at the end of a float glass production line with a throughput of around 2 million m² per annum. The temperature of glass leaving the float glass production line would be ideal for the closed space sublimation process and, with these production volumes, the cost per unit area would be small.

3.5.6 Integrally interconnected modules

One of the advantages of most thin-film solar cells is that the full module can be deposited with the cell interconnections made during deposition. The technique is illustrated in Fig. 3.42. The tin oxide coating on the glass is scribed, either mechanically or by laser, into a set of strips 1–2 cm wide running the full length of the module. The a-Si p-i-n junction or multijunctions are next deposited over the full area, and then scribed, with the scribe lines slightly offset from those of the tin oxide. The metal back contact is then deposited over the whole area, and then scribed with a further offset. Each individual cell is in the form of a long narrow strip, which reduces series resistance, and all cells are connected in series, with the outer metal strip being the negative terminal and the outer tin oxide strip the positive terminal for the whole module.

This technique reduces the handling required to a very considerable extent and reduces the manufacturing costs of thin-film modules quite significantly.

Summary

We have discussed the operation and manufacture of four types of thin-film cells which are in or near commercial production. All these cells have the potential for large-scale production at low unit cost. The reasons for the limited stability of amorphous silicon cells have been analysed, and a variety of structures under development which may overcome these problems have been

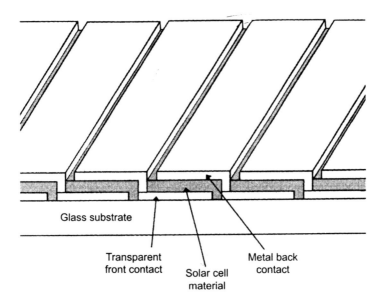

Fig. 3.42 Integrally interconnected module

discussed. We have shown that, with currently available technology and theoretical understanding, the thickness of polycrystalline silicon cells can be greatly reduced, leading to a promising new route for the production of efficient cells at a lower cost. We have also considered the structure and manufacture of two compound thin-film cells, CIS and CdTe. Finally, we have examined an important part of the thin-film photovoltaic technology: the manufacture of integrally interconnected modules. The fact that the manufacturing process produces modules rather than cells makes it much more amenable to large-scale production and lower cost. This has significant implications for the future of the PV industry.

SUMMARY OF THE CHAPTER

Solar cells convert sunlight into electricity by exploiting the photovoltaic effect. The maximum conversion efficiencies observed in laboratory devices have exceeded 30% but the typical efficiencies for production cells are 10–15%.

The physical principles of photovoltaic energy conversion have been discussed, starting from electronic properties of semiconductors. A variety of structures have been examined with a view of how power losses which accompany their operation can be reduced.

Crystalline silicon solar cells represent the largest part of the market. The technology has been reviewed in detail, starting from crystalline silicon production through wafer preparation and cell manufacture to the final module assembly. The traditional single-crystal technology, as well as the newer semicrystalline and ribbon techniques, have been examined. We have also taken a brief look at some devices in the research stage which are likely to shape the technology in the years to come.

Thin-film solar cells hold the promise of inexpensive technology with acceptable conversion efficiencies. The requirements on a thin-film cell have been considered. Four types of thin-film cells that are likely to be of commercial importance in the next few years have been examined in detail: the amorphous silicon cell, thin polycrystalline silicon cell grown on a low-cost substrate, the copper indium diselenide cell and the cadmium telluride cell.

BIBLIOGRAPHY AND REFERENCES

GREEN, M. A., *Solar Cells*, Prentice Hall, Englewood Cliffs, NJ, 1982.
HERSH, P. and ZWEIBEL, K., *Basic Photovoltaic Principles and Methods*, U.S. Government Printing Office, Washington, DC, SERI/SP-290-1448, 1982.
PULFREY, D. L., *Photovoltaic Power Generation*, Van Nostrand Rheinhold, New York, 1978.
VAN OVERSTRAETEN, R. and MERTENS, R., *Physics, Technology and Use of Photovoltaics*, Adam Hilger, Bristol, 1986.

SELF-ASSESSMENT QUESTIONS

PART A. True or false?

1. Majority carriers are the principal mobile charges created by light in semiconductors.

2. The conduction band is so-called because its structure determines the electrical properties of a semiconductor.

3. Boron contamination must be carefully avoided in silicon solar cell manufacture.

4. Cells made from low-bandgap semiconductors usually produce large current.

5. A solar cell can be pictured as a current generator connected in series with a diode.

6. The size of contacts in the point-contact cell aims to reduce the obscuration of the cell.

7. The main disadvantage of amorphous silicon for solar cell manufacture is its low light absorption.

8. The voltage output of a solar cell is directly proportional to the irradiance.

9. Recombination represents the principal voltage loss mechanism in good solar cells.

10. Series resistance mainly reduces the fill factor of a solar cell.

11. The front cell in a tandem must be made of high-bandgap semiconductor.

12. Light trapping is a means of increasing poor light absorption properties of most thin films.

13. Pure silicon is expensive because silicon ore is rare.

14. The slicing step is what limits the throughput of all ingot growth methods.

15. The major drawback of ribbon growth methods is their throughput.

16. The textured surface of a cell is rough as a result of the sawing step.

17. The reflectivity of bare silicon is high but decreases to acceptable levels after encapsulating the cells beneath glass.

18. The p-i-n structure of amorphous silicon cells is used for the beneficial effect of electric field in the region where carriers are generated.

19. CIS cells owe their importance to the fact that the thin film has good crystalline properties and no treatment of the grain boundaries is necessary.

20. The CdS layer in compound thin-film cells generates much of the current produced by the cell.

PART B

1. What are intrinsic, n-type and p-type semiconductors?

2. Define the efficiency of a solar cell. Under what conditions it is usually measured?

3. Write down the I–V characteristic of an ideal solar cell. How does it differ from the I–V characteristic of a diode?

4. Using Table 3.1, would you expect the open-circuit voltage of a crystalline silicon cell to be smaller or larger than for an amorphous silicon cell?

5. What are the principal current losses in a silicon solar cell?

6. What measures can be taken to minimise reflection from the top surface of the cell?

7. What is the highest efficiency attained by a silicon cell under standard test conditions?

8. List the five criteria which define the ideal thin-film cell.

9. Why does an amorphous silicon cell degrade under sunlight?

10. How can amorphous cells be made more efficient and more resistant to photodegradation?

11. How is light absorbed efficiently in thin polycrystalline silicon cells?

12. How is the performance of CIS cells improved by alloying with gallium?

13. What methods are used to deposit CdTe for solar cells?

14. What are the advantages of producing integrally interconnected modules?

PART C

1. Let us suppose that the photon spectral flux density in the solar spectrum can be approximated by the following function:

$$
\begin{aligned}
n(E) &= 0 & (E < a) \\
n(E) &= k_1(E - a) & (b > E \geq a) \\
n(E) &= k_2(c - E) & (c > E \geq b) \\
n(E) &= 0 & (E \geq c)
\end{aligned}
$$

where

$$
a = 0.4\,\text{eV}, \quad b = 0.85\,\text{eV}, \quad c = 3\,\text{eV},
$$

$$
k_1 = 7.55 \times 10^{17}\text{cm}^{-2}\text{s}^{-1}\text{ev}^{-2}, \quad k_2 = 1.58 \times 10^{17}\text{cm}^{-2}\text{s}^{-1}\text{ev}^{-2}.
$$

Plot the photon flux density and the corresponding solar spectrum.

2. (a) Using the spectrum of Example 1, assuming perfect absorption and maximum carrier collection at the junction, plot the electrical current that can be produced by a $10 \times 10\,\text{cm}$ solar cell made from a semiconductor with bandgap E_g. Based on this result, what current would you expect for (b) silicon cell; (c) gallium arsenide cell?

3. (a) Plot the efficiency of the solar cell in Example 1 as a function of the bandgap E_g of the semiconductor, assuming that the voltage produced by the cell is equal to $E_g/q - 0.4$ and the fill factor is 0.8.
   (b) What cell efficiencies would you predict for Si and GaAs solar cells?

Answers

Part A

1, False; 2, False; 3, False; 4, True; 5, False; 6, False; 7, False; 8, False; 9, True; 10, True; 11, True; 12, False; 13, False; 14, True; 15, False; 16, False; 18, True; 19, False; 20, False.

Part B

1. Intrinsic semiconductors are undoped; n-type and p-type contain donor and acceptor impurities, respectively. The electric current is carried predominantly by electrons in n-type, and by holes in p-type.
2. The efficiency of a solar cell in percent is the power output in milliwatts at the maximum power point under AM1.5 ($100\,\text{mW/cm}^2$) illumination, at $25\,^\circ\text{C}$.
3. $I = I_\ell - I_0[\exp(qV/kT) - 1]$. The diode characteristic does not contain I_ℓ.
4. Smaller since the bandgap of crystalline silicon is smaller.
5. Top-contact shading, top-surface reflection, collection losses by recombination, and incomplete absorption of light.
6. Antireflection coating (preferably in several layers) and/or surface texturing.
7. 23.3%
8. (a) Reasonable efficiency (over 10%). (b) Long lifetime (20 years or more). (c) Must not contain scarce materials. (d) Should produce integrally interconnected modules. (e) No environmental damage or hazard.
9. Light creates additional recombination or scattering centres (the Staebler–Wronski effect), thus reducing the cell output, particularly under bright sunlight.
10. Using structures formed by several thin layers based on a-S(H) or its alloys with carbon or germanium.
11. By exploiting the light-trapping effect, with reflecting back surface and top surface textured.
12. The alloy gives higher open-circuit voltage, better adhesion to the molybdenum contact, and lesser effect of light absorption in the window ZnO layer.
13. (a) Screen printing. (b) Spraying or other wet chemical methods. (c) Electroplating.
14. The module can be manufactured as a part of the cell production process, reducing the individual cell handling and interconnection and thus the total cost.

Part C

1. See Fig. E3.1.
2. See Fig. E3.2.
3. See Fig. E3.3. The model predicts identical efficiency of about 24% for both Si and GaAs cells. Other models often predict slightly higher efficiency for GaAs than for Si but, despite its simplicity, the model agrees remarkably well with the best efficiencies obtained for GaAs (25.1%) and Si (24.7%, see Sec. 3.4) solar cells to-date.

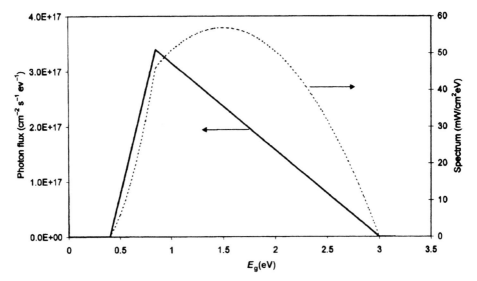

Fig. E3.1

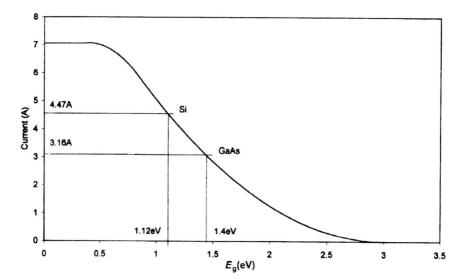

Fig. E3.2

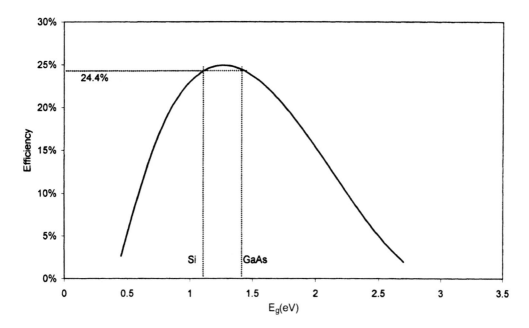

Fig. E3.3

4

Photovoltaic System Engineering

AIM

The aim of this chapter is to survey the structure, operation and design of photovoltaic systems

OBJECTIVES

On completion of this chapter, you should be able to:

1. understand the structure of a photovoltaic system, its subsystems, and the concept of sizing,
2. analyse the operation of the photovoltaic generator in terms of modules and their electrical characteristics,
3. examine the battery operation in photovoltaic systems,
4. evaluate the role of power conditioning and control elements,
5. size a simple system using the radiation power supply and load power requirements.

Edited from manuscripts of L. Castañer (Photovoltaic engineering), E. Lorenzo (Photovoltaic generator, Photovoltaic systems) and M. A. Egido (Sizing) Section 4.5.3 was written by J. N. Ross. Section 4.7 was written by C.V. Nayar, B. Wichert and W. B. Lawrance. Additional information by A. Sorokin and F. C. Treble.

NOTATION AND UNITS

Symbol		SI unit	Other unit
E_L	Daily load energy requirement	J	Wh
FF	Fill factor		
F	Fraction of daily load stored in battery		
G	Solar irradiance	W/m^2	
I_L	Average load current	A	
I_m	Module current at maximum power point	A	
I_{pv}	PV generator current under standard conditions	A	
I_{sc}	Short-circuit current	A	
k	Boltzmann constant	J/K	eV/K
LLP	Loss-of-load probability		
NOCT	Normal operating cell temperature	K	C
N_s	Number of series-connected modules		
N_p	Number of parallel strings		
n_c	Number of cells in a module		
P_{eff}	Nominal module power when connected to battery		
P_{max}	Power at maximum power point	W	
PSH	Peak Solar Hours		h
Q_{yd}	Yearly charge deficit	C	Ah
Q_{los}	Charge deficit to compensate for lack-of-sunshine days	C	Ah
Q_B	Nominal battery capacity	C	Ah
R	Load resistance	Ω	
R_s	Series resistance	Ω	
SF	Array oversize (safety) factor		
T_a	Ambient temperature	K	C
T_c	Cell operating temperature	K	C
t_i	Duration of operation of appliance i	s	h
V_{DC}	DC bus-bar voltage	V	
V_{oc}	Open-circuit voltage	V	
V_R	Voltage at resistive load	V	
V_m	Module voltage	V	
W_i	Nominal power rating of appliance i	W	
ΔE	Yearly energy deficit	J	Wh
Φ	Lowest permitted state of charge of the battery		
η_{bat}	Energy efficiency of the battery		

Unit conversion factors

$$\text{To convert} \begin{Bmatrix} \text{charge} \\ \text{energy} \\ \text{temperature} \end{Bmatrix} \text{from} \begin{Bmatrix} \text{Ah} \\ \text{Wh} \\ {}^\circ\text{C} \end{Bmatrix} \text{to} \begin{Bmatrix} \text{C} \\ \text{J} \\ \text{K} \end{Bmatrix} \begin{matrix} \text{divide by 3600} \\ \text{divide by 3600} \\ \text{add 273.16} \end{matrix}$$

Values of physical constants

$$k = 1.38 \times 10^{-23} \, \text{J/K} = 86.3 \times 10^{-3} \, \text{eV/K}$$

4.1 INTRODUCTION

Solar cells are not the only components of a PV system. Many other parts are usually required to provide a satisfactory electricity supply.

Many PV systems contain a provision for energy storage to supply electricity at night and during periods of inclement weather.

Solar cells generate direct current. Since most available appliances work with alternating current, some form of power conditioning is usually required.

Other power conditioning or control elements are needed to interface the different parts of the system, and to allow for the variable nature of the converted solar energy. Some elements also reflect the desirability to monitor the performance.

All these components have to be properly interconnected, sized and specified for PV operation.

In this unit, you will be given an overall view of the photovoltaic system, and how these problems are addressed in its design.

In section 4.2, we shall examine the structure of a system, and how the various parts, or subsystems, work together to provide the desired power output.

Each subsystem will then be analysed in more detail in sections 4.3–4.5. Their construction and components will be reviewed, and the main characteristics presented to you. You will be shown how solar cells, which were the subject of Chapter 3, can be integrated into a *photovoltaic generator*. We shall highlight the photovoltaic aspects of the storage and power conditioning subsystems which are also standard components of other energy storage and control systems.

The size of the system and, indeed, of the PV generator and storage subsystems, depend on the geographical location and on the application for which the system is intended. Solar energy can be a very reliable power source in isolated locations, with the minimum attention and maintenance. The reliability aspect of a PV system is therefore very important. As we shall see in this chapter, many decisions that go into the design of a system represent a careful consideration between the quest for maximum power production, and reliability. In particular, *sizing* of a photovoltaic system is an important part of its design, and you will be shown how it is done in section 4.6.

Many applications are best served by a combination of renewable and conventional energy sources to form a hybrid system. Such photovoltaic-diesel hybrid energy systems are discussed in section 4.7.

4.2 STRUCTURE OF A PHOTOVOLTAIC SYSTEM

The photovoltaic system consists of a number of parts or subsystems (Fig. 4.1):

(a) the photovoltaic generator with mechanical support and, possibly, a sun-tracking system,
(b) batteries (storage subsystem),

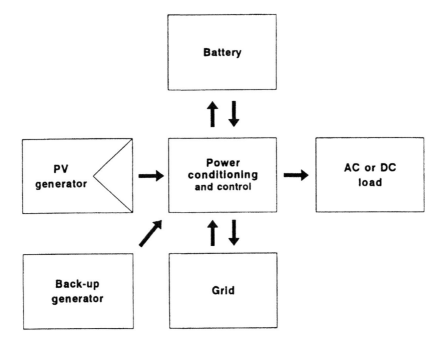

Fig. 4.1 The photovoltaic system. Usually, of course, not all the subsystems will be included

(c) power conditioning and control equipment, including provision for measurement and monitoring,

(d) the system may contain a supplementary or back-up generator (for example, a diesel generator) to form a hybrid system.

The choice of how and which of these components are integrated into the system is governed by various considerations.

There are two main categories of systems, *grid-connected* and *stand-alone*. The simplest form of the latter consists simply of a photovoltaic generator alone which supplies DC power to a load whenever there is adequate illumination. This type of system is common in pumping applications. In other instances, the system will usually contain a provision for energy storage by batteries. Some form of power conditioning is then frequently also included, as is the case when AC current is required at the output from the system. In some situations, the system contains a back-up generator.

Grid connected systems can be subdivided into those in which the grid merely acts as an auxiliary supply (*grid back-up*) and those in which it may also receive excess power from the PV generator (*grid interactive*)*. In PV power stations, all the generated power is fed into the grid.

* Or 'synchronously connected', in the language of electrical engineers.

Summary

We have reviewed the basic types of photovoltaic systems, and introduced the principal subsystems. Distinction has been made between grid-connected and stand-alone systems.

4.3 THE PHOTOVOLTAIC GENERATOR

4.3.1 Introduction

The heart of the system is the photovoltaic generator. It consists of *photovoltaic modules* which are interconnected to form a DC power-producing unit. The physical assembly of modules with supports is usually called an *array* (Fig. 4.2).

The module construction was introduced in Chapter 3. In section 4.3.2 of this chapter, we examine the electrical characteristics of a module in detail and show how to determine the module output during practical operation.

Module interconnection in a photovoltaic generator will be discussed in section 4.3.3 where we describe the characteristics of a photovoltaic generator under conditions that are encountered in practical operation.

In section 4.3.4, we discuss the two principal options for panel orientation: fixed inclination, or tracking.

4.3.2 Photovoltaic modules

The photovoltaic module was already encountered in Chapter 3. As the name suggests, the module represents the basic construction unit of a photovoltaic generator. The structure of a module based on crystalline or semi-crystalline silicon cells (discussed in section 3.4.5) is shown in Fig. 4.3. It is this type of

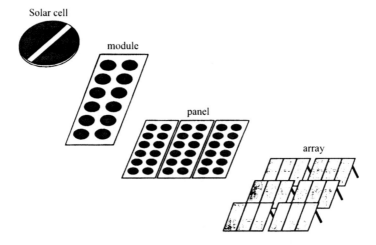

Fig. 4.2 The photovoltaic hierarchy

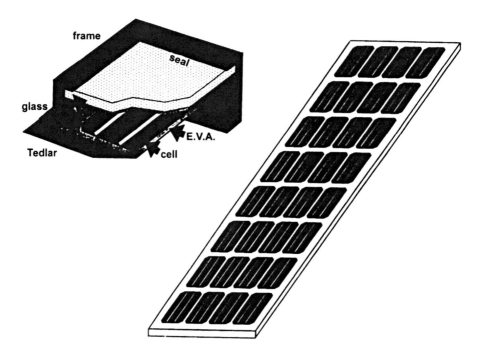

Fig. 4.3 The photovoltaic module based on crystalline silicon solar cells

module that is most commonly used in practice, and we shall consider its electrical characteristics in more detail.

Most frequently, the cells in a module are interconnected in series. The reason comes from the electrical characteristics of an individual solar cell As we have seen in Chapter 3, a typical 4-inch diameter crystalline silicon solar cell, or a 10 cm × 10 cm multicrystalline cell, will provide between 1 and 1.5 W under standard conditions, depending on the cell efficiency. This power is usually supplied at a voltage 0.5 to 0.6 V. Since there are very few appliances that work at this voltage, the immediate solution is to connect the solar cells in series.

The number of cells in a module is governed by the voltage of the module. The nominal operating voltage of the system usually has to be matched to the nominal voltage of the storage subsystem. Most of photovoltaic module manufacturers therefore have standard configurations which can work with 12 V batteries. Allowing for some overvoltage to charge the battery and to compensate for lower output under less-than-perfect conditions, it is found that 36 solar cells in series ensures reliable operation, as discussed below on the example of the module power output. In self-regulating systems (section 4.5.2), the module voltage is required to control the charging current of the battery, and the number of cells is usually smaller (often 32 to 34).

The power of silicon modules thus usually falls between 40 and 60 W. The module parameters are specified by the manufacturer under the following standard conditions:

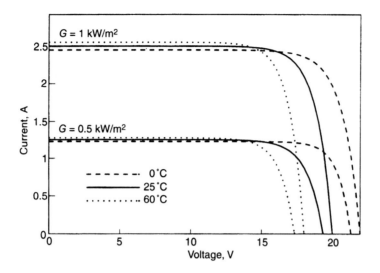

Fig. 4.4 The temperature and irradiance dependence of the module I–V characteristic

$$
\begin{array}{lll}
\text{Irradiance} & 1\,\mathrm{kW/m^2} & \\
\text{Spectral distribution} & \text{AM1.5} & (4.1) \\
\text{Cell temperature} & 25\,^\circ\mathrm{C} &
\end{array}
$$

Indeed, they are the same conditions as are used to characterise solar cells. The nominal output is usually called the *peak power* of a module, and expressed in *peak watts*, W_p.

The three most important electrical characteristics of a module are the short-circuit current, open-circuit voltage and the maximum power point as functions of the temperature and irradiance. These characteristics resemble the I–V characteristic of a solar cell (Fig. 4.4) but some specific features need to be highlighted.

Temperature is an important parameter of a PV system operation. As we have seen in Chapter 3, the temperature coefficient for the open-circuit voltage is approximately equal to $-2.3\,\mathrm{mV/^\circ C}$ for an individual cell. The voltage coefficient of a module is therefore negative and large since some 36 cells are connected in series. The current coefficient, on the other hand, is positive and small—about $+6\mu\mathrm{A/^\circ C}$ for a square centimetre of the module area. Accordingly, only the voltage variation with temperature is allowed for in practical calculations, and for an individual module consisting of n_c cells connected in series is set equal to

$$
\mathrm{d}V_{oc}/\mathrm{d}T = -2.3 \times n_c \ \mathrm{mV/^\circ C} \qquad (4.2)
$$

It is important to note that the voltage is determined by the operating temperature of the cells which differs from the ambient temperature (see equation (4.5) below).

As for a single cell, the short-circuit current I_{sc} of a module is proportional to the *irradiance*, and will therefore vary during the day in the same manner. Since the voltage is a logarithmic function of the current, it will also depend logarithmically on the irradiance. During the day, the voltage will therefore vary less than the current. In the design of the PV generator, it is customary to neglect the voltage variation, and to set the short-circuit current proportional to the irradiance:

$$I_{sc}(G) = I_{sc} \; (\text{at} 1 \, \text{kW/m}^2) \times G \; (\text{in} \; \text{kW/m}^2) \qquad (4.3)$$

The operation of the module should lie as close as possible to the *maximum power point*. It is a significant feature of the module characteristic that the voltage of the maximum power point, V_m, is roughly independent of irradiance. The average value of this voltage during the day can be estimated as 80% of the open-circuit voltage under standard irradiance conditions. This property is useful for the design of the power conditioning equipment.

The characterisation of the PV module is completed by measuring the *Normal Operating Cell Temperature* (NOCT) defined as the cell temperature when the module operates under the following conditions at open circuit:

Irradiance	$0.8 \, \text{kWm}^2$	
Spectral distribution	AM1.5	
Ambient temperature	$20 \, ^\circ\text{C}$	(4.4)
Wind speed	$> 1 \, \text{m/s}$	

NOCT (usually between $42 \, ^\circ\text{C}$ and $46 \, ^\circ\text{C}$) is then used to determine the solar cell temperature T_c during module operation. It is usually assumed that the difference between T_c and the ambient temperature T_a depends linearly on the irradiance G in the following manner:

$$T_c - T_a = \frac{\text{NOCT} - 20}{0.8} G(\text{kW/m}^2) \qquad (4.5)$$

The output power of a photovoltaic module depends on the system configuration and, in particular, on the method which is used to control the operating DC voltage. Small to medium stand-alone systems (with peak power of less than about 1kW_p) rarely employ charge regulators with maximum power point tracking and the module voltage is set largely by the battery. The power characteristic of a module connected to a battery with nominal voltage of 12V is shown in Fig. 4.5. It is seen that, under most circumstances, the module will operate in the linear part of its $I\text{–}V$ characteristic supplying the same current I_{sc} as at short circuit. The output power is then given by

$$P = I_{sc}(G) V_{bat} = G P_{eff}$$

where, using (4.3)

$$P_{eff} = V_{bat} I_{sc}$$

Figure 4.6 summarises the calculation of module parameters under operating conditions in a system using a MPP tracker. The short-circuit current is

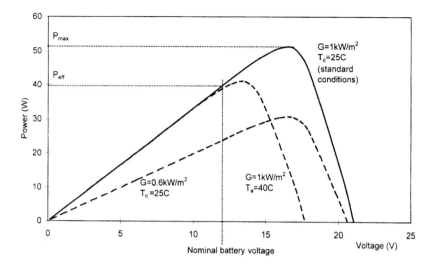

Fig. 4.5 Voltage dependence of the power produced by a PV module as a function of irradiance and cell or ambient temperature T_c or T_a

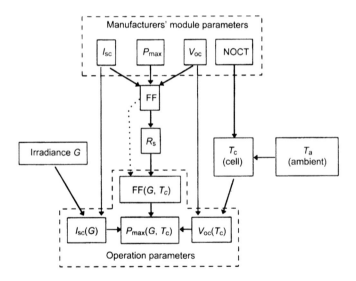

Fig. 4.6 Calculation of the module operation parameters in systems using MPP tracker

determined from equation (4.3). The cell operating temperature is found using equation (4.5) and the resulting value is then used to obtain the open-circuit voltage by using (4.2). The temperature and irradiance dependence of the fill factor is complicated, but can be determined by calculating first the series resistance which is independent of the module conditions. In a simplified calculation, the fill factor can be assumed constant. These parameters then yield the maximum power of the module under operating conditions.

4.3.2.1 Worked example

Determine the parameters of a module formed by 34 solar cells in series, under the operating conditions $G = 700 \, \text{W/m}$, and $T_a = 34\,°C$. The manufacturer's values under standard conditions are: $I_{sc} = 3\,\text{A}$; $V_{oc} = 20.4\,\text{V}$; $P_{max} = 45.9\,\text{W}$; NOCT $= 43\,°C$

Solution

1. Short-circuit current
   $I_{sc} = (700\text{W/m}^2) = 3 \times 0.7(\text{kW/m}^2) = 2.1\,\text{A}$.

2. Solar cell temperature
   $T_c = 34 + 0.7 \times (43 - 20)/0.8 = 54.12\,°C$.

3. Open-circuit voltage
   $V_{oc}(54.12\,°C = 20.4 - 0.0023 \times 34 \times (54.12 - 25) = 18.1\,\text{V}$

4. We shall now determine the maximum power point using the simplifying assumption that the fill factor is independent of the temperature and the irradiance:
   FF $= 45.9/(3 \times 20.4) = 0.75$
   $P_{max}(G, T_c) = 2.1 \times 18.1 \times 0.75 = 28.5\,\text{W}$.

   Thus, noting the manufacturer's value of P_{max} we see that the module will operate at about 62% of its nominal rating.

4.3.3 Interconnection of PV modules

A schematic diagram of a PV generator consisting of several modules is shown in Fig. 4.7. In addition to photovoltaic modules, the generator contains bypass and blocking diodes. As we shall see below and in section 4.4, these diodes protect the modules and prevent the generator acting as a load in the dark.

The modules are connected in series to form strings, where the number of modules N_s is determined by the selected DC bus voltage, and the number of parallel strings N_p is given by the current required from the generator. For the generator in Fig. 4.7, $N_s = 2$ and $N_p = 3$. The output voltage would therefore be twice the module voltage, and the current three times the module current. We shall discuss these points further in section 4.6.

This analysis assumes that all the modules are identical. In practice, the modules (and cells) are not identical, and their parameters exhibit a certain degree of variability for two principal reasons.

- The solar cells and modules vary in quality as a result of the manufacturing process. In general, the current produced by commercial modules suffers a higher degree of dispersion than the voltage.

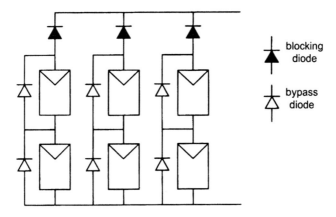

Fig. 4.7 The photovoltaic generator

- Different operating conditions may exist in different parts of the PV array. For example, one must allow for different cleanness of different parts of the PV generator, or some modules may be obscured by a cloud which is covering only a part of the array.

This variability of component parameters has two important effects. Firstly, the output power of the generator is less than the sum of the values corresponding to all the constituent modules. This gives rise to *mismatch losses*. These losses can be minimised by sorting the modules prior to interconnection, and forming series strings from modules with similar values of short-circuit current.

Secondly, there is a potential for overheating the 'poorest' cell of a series string. In some circumstances, a cell can operate as 'load' for other cells acting as 'generators'. Consequently, this cell dissipates energy and its temperature increases. If the cell temperature rises above a certain limit (85–100°C) the encapsulating materials can be damaged, and this will degrade the performance of the entire module. This is called the *hot-spot* formation.

This effect is illustrated in Fig. 4.8 which shows a cell in a string which does not produce current—this can happen, for example, when the cell is shaded. We learned in Chapter 3 that a cell in the dark acts as a diode. The shading of one cell—converting it into a diode under reverse bias—therefore eliminates the

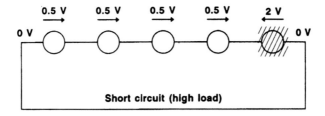

Fig. 4.8 The hot-spot formation

current produced by the entire string. Furthermore, the shaded cell will dissipate all the power produced by the illuminated cells in the string which can be considerable if the string is large.

The common technique used to alleviate this effect is to employ *bypass diodes* which are connected across a block of several cells in a string (Fig. 4.7). This limits the power which is dissipated in this block and provides a low-resistance path for the module current.

4.3.4 Capturing the sunlight

An important problem that confronts the designer of an array is whether the modules are to be mounted at fixed positions, or their orientations will follow ('track') the motion of the Sun.

In most arrays, the modules are supported at a fixed inclination facing the equator. This has the virtue of simplicity—no moving parts and low cost. As we have seen in Chapter 2, the optimum angle of inclination depends mainly on the latitude, the proportion of diffuse radiation at the site, and the load profile.

By mounting the array on a two-axis tracker, up to 40% more of the solar energy can be collected over the year as compared with a fixed-tilt installation. But this increases complexity and results in lower reliability and higher maintenance costs. Single-axis tracking is less complex but yields a smaller gain. Where labour is available, the orientation may be manually adjusted to increase the output. It has been estimated that, in sunny climates, a flat plate array moved to face the sun twice a day and given a quarterly tilt adjustment can intercept nearly 95% of the energy collected with a fully automatic two-axis tracking.*

Tracking is particularly important in systems which operate under concentrated sunlight. The structure of these systems ranges from a simple design based on side booster mirrors, to concentration systems which employ sophisticated optical techniques to increase the light input to the cell by several orders of magnitude. These systems must make allowance for an important fact that concentrating the sunlight reduces the angular range of rays that the system can accept for conversion. Tracking becomes necessary once the concentration ratio exceeds about 10 and the system can only convert the direct component of solar radiation.

In this chapter, we shall follow the established practice and lay emphasis on systems with flat panels positioned at a fixed orientation. Concentration systems will be discussed in detail in Chapter 7.

Summary

The structure and properties of the photovoltaic generator and its basic building block, the module, have been examined. The electrical characteristics of the

---

* F. C. Treble, lecture notes of the Southampton Short Course on Solar Energy Conversion and Applications.

generator depend on the ambient temperature and level of irradiation, and we have shown how the output from a PV generator under operating conditions can be determined. The mismatch losses, and measures that can be taken to minimise them, have been discussed. The relative merits of the main options regarding the panel orientation—fixed inclination or tracking—have also been considered.

4.4 ENERGY STORAGE

4.4.1 Introduction

Since the solar energy supply is intrinsically variable in time, stand-alone photovoltaic systems usually make a provision for energy storage.

Although a variety of energy storage methods are under consideration (Table 4.1), the majority of stand-alone PV systems today use battery storage. The notable exception are water pumping systems where the storage of pumped water is the preferred alternative (see section 5.4). The batteries in most common use are lead-acid batteries because of their good availability and cost effectiveness. Their specific features will be discussed below. Nickel cadmium batteries are used in some smaller applications where their ruggedness, both mechanical and electrical, is considered essential. However, their high cost per amount of energy stored has prevented their wider use in photovoltaics.

As we shall see in Chapter 5, future use of PV power to supplement grid electricity is being seriously addressed in many industrialised countries. The grid would then provide the natural and cheapest storage option. In the longer term, hydrogen production, combined with PV power generation, offers the attractive possibility of comprehensive energy technology with minimum environmental risks, as discussed in Chapter 7.

Table 4.1 Some energy storage systems

Energy stored	Technology	Remarks
	Pumped water	1. Common utility use as large-scale energy storage 2. PV pumping (Chapter 5)
Mechanical	Compressed air	Demonstrated technology for large-scale storage
	Flywheel	Under investigation for small systems
Electromagnetic	Electric current in superconducting ring	New development potential using 'high-temperature' super conducting materials
Chemical	Batteries	Discussed in this section
	Hydrogen production	See Chapter 7

4.4.2 Battery operation in PV systems

Batteries in PV systems operate under specific conditions which must be allowed for in the system design, as they affect both battery life and the efficiency of the battery operation. The most prominent feature is *cycling* with various cycles of different degree of regularity (Fig. 4.9). During the *daily cycle*, the battery is charged during the day and discharged by the night-time load. The depth of discharge in the daily cycle for systems without back-up is always fairly shallow.

Superimposed on the daily cycle is the *climatic cycle*, due to the variable climatic conditions. This cycle occurs anytime when the daily load exceeds the average energy supply from the PV generator.

In systems where reliability is not the paramount consideration, the battery may act as a *seasonal buffer*. In this instance, the climatic cycle extends over a substantial part of the season (Fig. 4.10).

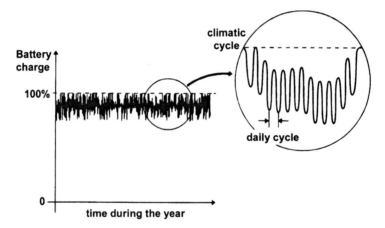

Fig. 4.9 The cyclic operation of battery in PV systems

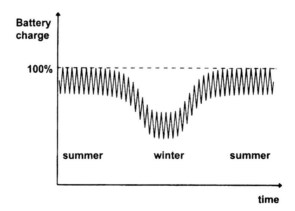

Fig. 4.10 The seasonal cycle

The details of the cyclic operation depend on the desired reliability of the system, and on the relationship between the storage capacity and size of the PV generator. We return to these points in section 4.6 when we discuss sizing.

4.4.3 Lead-acid batteries

In simplified terms, the lead-acid battery comprises two electrodes of lead and lead dioxide, and the electrolyte of sulphuric acid diluted with water. In practical construction, the electrodes are formed by a lead grid (sometimes alloyed with calcium or antimony) carrying the active material in the form of a porous structure that offers a large surface area for chemical reactions with the electrolyte.

The chemical reactions that take place during battery operation are shown in Fig. 4.11. During the charging process, lead oxide is formed at the anode, pure lead is formed at the cathode, and sulphuric acid is liberated in the electrolyte. During discharge, lead sulphate is formed at both electrodes and sulphuric acid is removed from the electrolyte.

A typical voltage behaviour during the discharge of a lead-acid battery is shown in Fig. 4.12. It is seen that the battery capacity decreases markedly when discharged at a high rate. For example, a battery with nominal capacity specified at the 10 h discharge rate can markedly increase its capacity under 100 h discharge—the rating value which is used in many photovoltaic applications. This characteristic must be taken into account when designing the energy storage of a PV system since many operate under different conditions than those specified by the battery manufacturers.

The voltage behaviour during charging is shown in Fig. 4.13. After a relatively slow increase up to about 2.35 V per element, there is a steep voltage rise

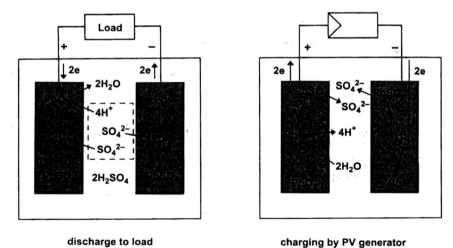

discharge to load   charging by PV generator

Fig. 4.11 Lead-acid battery operation

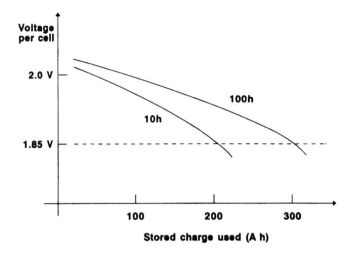

Fig. 4.12 Battery discharge characteristics under different discharge rates

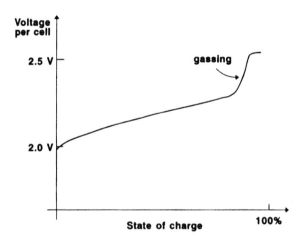

Fig. 4.13 Battery charging characteristic

accompanied by *gassing*—the generation of hydrogen and oxygen at the electrodes.

These chemical reactions imply various deleterious processes that accompany the battery operation and shorten battery life or increase maintenance requirements.

Gassing increases the need for maintenance and may represent a safety hazard. In moderate levels, it can be used to advantage by alleviating stratification of the electrolyte (see below).

The repeated growth and dissolution of lead, lead oxide and lead sulphate with different specific volumes is accompanied by mechanical stresses at the electrodes. These stresses, in turn, result in *shedding of active matter* from the electrodes. Another instance where this phenomenon may occur is during intense gassing.

Sulphation is the formation of large lead sulphate crystals at the plates which hinder the reversible chemical reactions. This phenomenon occurs mainly when the battery remains in a low-charge state for extended periods of time.

The battery operation tends to favour a non-uniform electrolyte distribution where the electrolyte with the highest density occurs at the bottom of the battery vessel. This *stratification* of the electrolyte promotes corrosion and sulphation of the bottom part of the negative electrode, but can be avoided by a regular weak overcharge in which gassing is used to stir the electrolyte.

It is a good practice to give the batteries an *equalising charge* at the end of a period when the battery will have operated in low-charge state (normally, at the end of winter). This boost charge which briefly over-charges the battery ensures that all the elements are fully recharged.

Electrode corrosion occurs particularly by the growth of the positive lead grid, and is accelerated by elevated temperatures.

The design of the battery storage must also ensure that the electrolyte does not freeze during operation at sites where the temperature can drop to low levels. This can be avoided by maintaining a relatively high density of the electrolyte, either by high state of charge or by increasing the electrolyte density above the usual norm.

Cycling characteristics have a pronounced effect on the lifetime of lead-acid batteries, and this should be taken into account in the system design. Some manufacturers provide an indication of the battery life as a function of the number of cycles and the depth of discharge (Fig. 4.14). The system design will

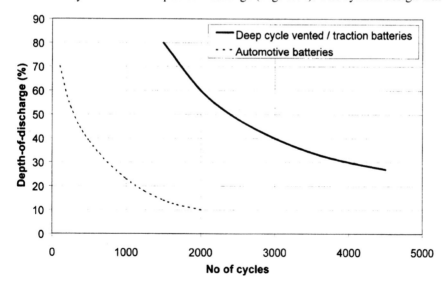

Fig. 4.14 An example of a manufacturer's specification of battery lifetime as a function of the number of cycles and depth of discharge

Table 4.2 Principal types of commercial lead-acid batteries

Type	Application	Operation	Characteristics
SLI	Vehicle starter batteries	Supply of high peak power for short time	Low resistance to cycling
			Low cost
Traction	Electric vehicles	Deep and frequent cycles	High resistance to cycling
			High water consumption
			Frequent maintenance
Stationary storage	Emergency power supplies	Float operation in high state of charge	Moderate resistance to cycling
			Low water consumption

need to take into account the efficiency of energy storage by the battery. The energy returned upon discharge is lower than the energy supplied upon charging because the battery voltage is higher during charging than during discharge. Some charge may also be lost in the battery, principally during gassing. The overall energy efficiency of most lead-acid batteries is likely to be in the range between 85 and 90%.

Although specialised PV batteries are now becoming available on the market, most batteries that are currently installed in PV systems are standard components originally intended for conventional application, or adapted from them to suit the particular mode of operation envisaged for the PV system. The three principal classes of lead-acid batteries are listed in Table 4.2.

Summary

Lead-acid batteries provide the most common means of energy storage in PV systems today. A prominent feature of their operation is cycling. This, together with other operation parameters, affects the battery life and maintenance requirements which must be allowed for in the design of a PV system.

4.5 POWER CONDITIONING AND CONTROL

Various electronic devices are used to accomodate the variable nature of power output from the PV generator, to avoid the malfunction of the system, or to convert the DC power produced by the PV generator into AC output.

4.5.1 The blocking diode

We have seen in Chapter 3 that a solar cell in the dark behaves as a diode. Similarly, the characteristic of a photovoltaic generator at night can be

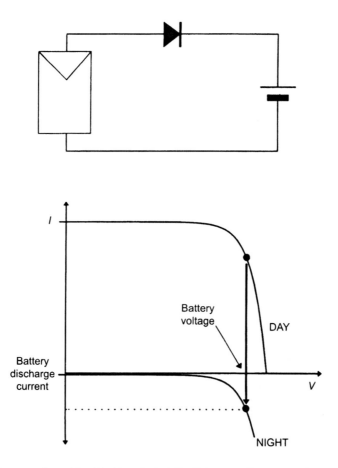

Fig. 4.15 The operation of the blocking diode (after E. Lorenzo, Electricided Solar Fotovoltaic, Publication of E.T.S.I. Telecommunication, Madrid, 1984)

obtained by displacing its usual characteristics under illumination along the current axis, until it passes through the origin. Without special precautions, this type of night-time operation of the photovoltaic generator will provide a discharge path for the battery. The simplest solution is to separate the generator and battery by a *blocking diode* (Fig. 4.15). When the voltage at the battery exceeds the voltage at the generator, the diode becomes reverse-biased and prevents the battery discharge. During daytime operation, however, there will be a voltage drop across the blocking diode which should be taken into account when designing the system.

In systems using modern PV modules where the series resistance is low and the *I–V* characteristic approaches the ideal curve, the battery discharge current via the PV generator at night can be very small. The power dissipated at the blocking diode during daytime operation may then exceed the night-time discharge losses. For this reason, the blocking diode is sometimes omitted from the circuit design.

4.5.2 Self-regulating systems

In small systems, the connection of the battery with the PV generator via a blocking diode provides the simplest solution. This configuration relies on a correct choice of the operating point of the PV generator to match the battery charging requirements (Fig. 4.16). The usual operation of the system takes place between the battery voltage under the minimum allowed state of charge, and the voltage at full charge, allowing for a voltage drop across the blocking diode. If the array operating voltage is set at the upper end of this voltage range, a slight increase in the battery voltage then sharply reduces the charging current from the PV generator and prevents battery overcharging.

We should emphasise, however, that the simplicity of design in these systems suffers serious drawbacks in practical operation. The fact that there is no battery discharge protection can have a very detrimental effect on the battery life. The system design is very sensitive to the ambient temperature and, in many circumstances, would yield very inefficient performance. For these reasons, self-regulation is usually recommended only in small systems, and when system costs must be kept at the absolute minimum.

4.5.3 Charge regulator

As discussed in section 4.4, measures must be taken to prevent excessive discharge or overcharge of batteries. This is the function of the charge regulator.

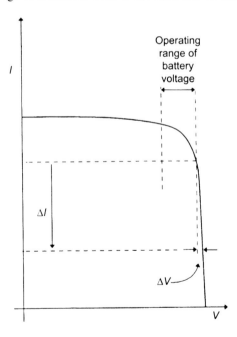

Fig. 4.16 The operation of self-regulating PV system

Excessive discharge is avoided by monitoring the battery voltage and disconnecting the load from the battery if the voltage falls below a pre-set minimum value. This would typically be about 10.8 V per cell for a lead-acid battery. The regulator will not reconnect the load until the battery voltage has risen to a value significantly higher than this minimum value. This is necessary to ensure that the load is not reconnected until some charge has been returned to the battery.

While power is being supplied by the PV array the main role of the regulator is to limit the maximum battery voltage to prevent excessive gassing and to prevent overcharging the battery. Unless a very large array is used with a very small battery, it is not normally necessary to limit the peak current during the *bulk charge* phase, when the battery voltage is below the limiting value. The voltage regulation may be achieved by the use of either a *shunt regulator* or a *series regulator*.

A shunt regulator has a variable resistance element in parallel with the battery. As the resistance is reduced more of the current from the PV array is diverted through the resistor and less through the battery. The variable resistance element will generally be a transistor (bipolar or MOSFET) (Fig. 4.17(a)). The disadvantage of the shunt regulator is that it may dissipate a large amount of power.

An alternative is to use a series regulator (Fig. 4.17(b)), in which a variable resistance element is included in series with the PV array and the battery. As the battery is charged and its voltage rises the series resistance is increased

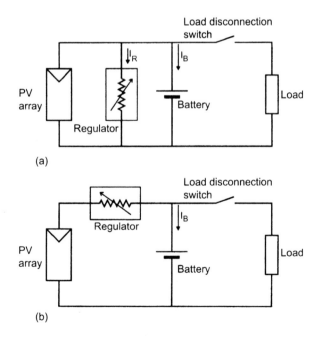

Fig. 4.17 (a) Shunt regulator; (b) series regulator

to reduce the battery voltage (and current). This arrangement dissipates little power in the resistive element before the regulator starts to operate (the voltage across the element is small), and little power when the battery is fully charged (the current is small). The series regulator thus dissipates much less power than the shunt regulator, provided the array open-circuit voltage is chosen correctly.

The power dissipated in the series regulator can be greatly reduced by using a switch as the series element. The switch could be mechanical, but more usually a MOSFET or bipolar transistor would be used, with a control to turn the transistor hard on, or off. Various control strategies are possible using a series switch to control the maximum charging voltage. These strategies rely on the property of the battery that when the charging current is disconnected the battery voltage falls back quite slowly to its steady-state, open-circuit value, and rises quite slowly when reconnected. By setting the switch control circuit to open the switch at one battery voltage and close it again at a slightly lower value, the voltage can be held within limits and the mean current reduced by a crude form of pulse width modulation (PWM – see also Box 4.1 Switching Converters and Inverters). This is illustrated in Fig. 4.18. As the battery approaches a full state of charge the fraction of the time for which the switch is on reduces, reducing the mean charging current.

More sophisticated controllers will allow the battery voltage to rise to a higher limit value after a deep or prolonged discharge, and then reduce the upper voltage limit to avoid overcharging the battery. This helps return the battery charge rapidly to a high state and also to allow the battery to 'gas', stirring up the electrolyte. The limit values may also be varied to allow for the effects of temperature variation.

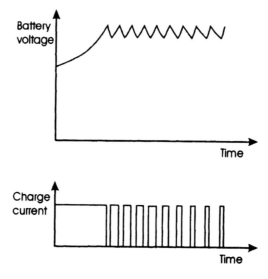

Fig. 4.18 Charge control with simple switching charge regulator

Box 4.1 Switching power converters and inverters

DC/DC converters

Switching power converters are widely used to transform DC power between one voltage and another. There are a wide variety of circuits used. Figures B4.1 and B4.2 illustrate the basic circuit topology of two of the simplest types of converter. The buck converter (Fig. B4.1) reduces the voltage while the boost converter (Fig. B4.2) increases it. In both cases the voltage transformation is performed with only a small loss of power. The switch S is an electronic switch, usually a field effect transistor (MOSFET) or, at higher power levels, an insulated gate bipolar transistor (IGBT). This switch is able to switch on and off at high speed, with low resistance when on and very high resistance when off. The switches are generally driven at a constant switching frequency, and the ratio of the time for which the switch is on to the period (the duty ratio D) is varied to control the load voltage. This is referred to as pulse width modulation (PWM). Less commonly the pulse frequency may be varied to change the duty ratio. This is referred to as pulse frequency modulation (PFM).

In both circuits the current flowing in the inductor rises while the switch is on, storing energy in the magnetic field of the inductor. The rate of rise of

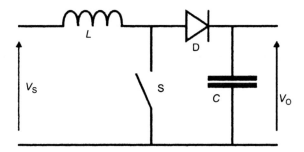

Fig. B4.1 Circuit diagram of the buck converter

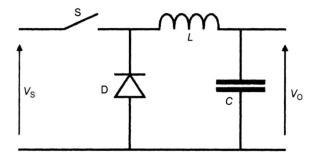

Fig. B4.2 Circuit diagram of the boost converter

current in the inductor is proportional to the voltage across it. When the switch is turned off the current in the inductor must continue to flow. The voltage across the inductor changes sign rapidly and the current is diverted to flow through the diode. The current in the conductor then falls until either it reaches zero (discontinuous current operation), or until the switch turns on again (continuous current operation).

Either of these converters will operate in discontinuous current mode at low load current or continuous current mode at high load current. The load current at the boundary between these two regions of operation is determined by the value of the inductance, the switching frequency and the source voltage. In continuous current mode, the ratio of the load voltage, V_o, to the source voltage, V_s, is determined only by the duty ratio, D. For a buck converter the ratio is given by,

$$\frac{V_o}{V_s} = \frac{1}{1 - D}$$

while, for the boost converter,

$$\frac{V_o}{V_s} = D$$

The expressions for the voltage ratios are more complex for discontinuous current operation, and depend on the load current. Converters are frequently designed to operate in continuous current mode at high load current in order to minimise the peak current in the switches and other components. The control system must, however, be carefully designed to ensure stable operation over the whole range of load current.

When used for power point tracking the duty ratio must be controlled to keep the solar array at its optimum voltage. This may be achieved in several ways. One method of finding the optimum conversion ratio is to periodically vary the duty ratio. This will modulate the voltage across the solar array, and hence the array current and power. By observing if an increase in duty ratio increases or decreases the array power the mean duty ratio may be varied so as to maximise the power. An alternative, possibly simpler, approach is to periodically disconnect the array for a very short interval so that the open circuit voltage may be measured. As we have seen in section 4.3.2, the maximum power point voltage of a PV array is about 80% of the open circuit voltage. By adjusting the duty ratio so as to make the on-load array voltage equal to this optimum value, the power transfer may be optimised.

Self-commutating inverters

Inverters convert power from DC to AC. They do this by using electronic switches to reverse the polarity of the electricity supplied to the load periodically. The term *self-commutating* refers to the use of transistor switches to reverse the polarity of the supply. These switches are able to turn on and off under internal control. This distinguishes the self-commutating inverter from the *line commutated* inverter which uses thyristor switches and requires an alternating load voltage from an external source periodically to turn the

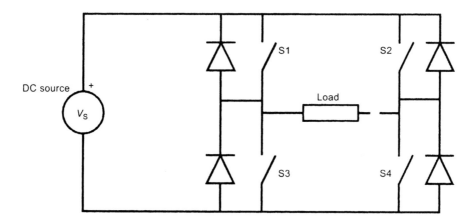

Fig. B4.3 A single-phase full bridge inverter

switches off. Self-commutated converters are generally preferred at low to moderate power levels (up to tens of kW) while line commutated inverters may operate at very much higher power levels.

Low-power, single-phase inverters generally use four controlled switches in a bridge arrangement as shown in Fig. B4.3. The switches will generally be insulated gate bipolar transistors (IGBTs) which are able to switch rapidly and operate at voltages of up to 1000V. Operated in the simplest manner the inverter will generate a square wave by closing S1 and S4 while opening S2 and S3 for one half cycle, and then closing S2 and S3 and opening S1 and S4 for the second half cycle. The diodes are necessary to provide a path for load current to flow when both S1 and S2, or S3 and S4, are open simultaneously, as they must be briefly during the changeover.

One serious disadvantage of this simple inverter is that the square voltage waveform, which is rich in odd order harmonics may not be suitable for supplying many AC appliances. Filtering out these frequency components involves large and costly filters. The second disadvantage is that there is no control of the load voltage which is determined by the source voltage. Both these disadvantages may be overcome by the use of sinusoidal pulse width modulation. This involves a very similar concept to that used in the buck converter. The switches are operated at a high frequency, typically greater than 20 kHz and the duty ratio is varied in such a way that the load voltage (averaged over the switching period) varies sinusoidally. The high-frequency components of the load voltage associated with the switching are easily filtered out with a small and cheap filter. An example of the resulting PWM voltage waveform is shown in Fig. B4.4, together with the sinusoidal wave corresponding to the fundamental component of the output voltage. The sinusoidal frequency is 50 Hz, and in this case an unrealistically low switching frequency of 1 kHz has been chosen so that the shape of the pulses may be seen. The corresponding frequency spectrum is shown in Fig. B4.5. Clearly, the fundamental (50 Hz) component is well separated from the switching frequency components, facilitating their removal by a suitable filter. By

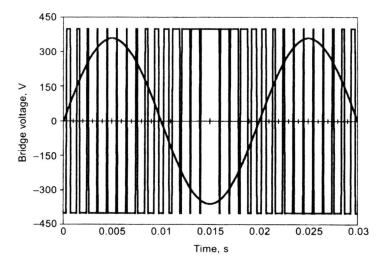

Fig. B4.4 The output voltage of a pulse width modulated inverter, also showing the corresponding fundamental component of the output voltage

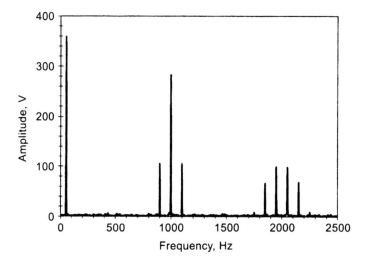

Fig. B4.5 Frequency spectrum of PWM inverter

controlling the maximum and minimum duty ratios the maximum and minimum values of the sinusoidal.

The main disadvantage of PWM inverters is that like all switching converters they generate electrical noise at frequencies extending up to frequencies of several MHz. Careful filtering and screening is necessary to meet the electro-magnetic compatibility (EMC) regulations.

(Written by J. N. Ross)

4.5.4 DC/DC converter

The variability of the power output from the PV generator implies that, without special interface measures, the generator will often operate away from its maximum power point. The associated losses can be avoided by the use of a *maximum-power-point tracker* which ensures that there is always a maximum energy transfer from the generator to the battery.

The principles of the MPP tracker are demonstrated in Fig. 4.19 for the situation when the PV generator feeds power to a resistive load R. Shown are the $I-V$ characteristics of the generator and the load, together with constant power curves $P = IV =$ constant. It is seen that at the operating point (which corresponds to a direct connection between the generator and the load) the delivered power is significantly below P_{max}, the maximum power of the PV generator. A DC/DC converter will transform the voltage at the load to a value

$$V_R = \sqrt{P_{max}R} \qquad (4.6)$$

thus ensuring a maximum power transfer.

The use of a MPP tracker is usually justified in systems with peak power in excess of $1\,kW_p$. The principles of operation of DC/DC converters are discussed further in Box 4.1.

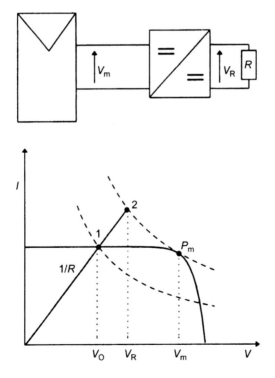

Fig. 4.19 The operation of MPP tracker (after E. Lorenzo, Electricidad Solar Fotovoltaica, Publication of E.T.S.I. Telecommunicacion, Madrid, 1984)

4.5.5 DC/AC converter (inverter)

This is a standard item of electronic equipment which is used in many different applications. The input power is the DC power from the photovoltaic generator or battery, and the output is AC power used to run AC appliances or fed into the utility grid. The efficiency of the inverters usually depends on the load current being a maximum at the nominal output power. It can be as high as 95% but will be lower (75–80%) if the inverter runs under part load.

The majority of inverters for PV applications can be classified into three main categories (Fig. 4.20). *Variable frequency inverters* are used for stand-alone drive/shaft power applications, almost exclusively in PV pumping systems. The remaining two main inverter categories are suitable for the grid connection of PV power plants. *Self-commutating fixed-frequency inverters* are able to feed an isolated distribution grid and, if equipped with special paralleling control, also a grid supplied by other parallel power sources. *Line-commutated fixed-frequency inverters* are able to feed the grid only where the grid frequency is defined by another power source connected in parallel. The inverter will not work if such external frequency reference is lacking (e.g. during grid black-out). Further discussion of inverters can be found in Box 4.1.

4.5.6 Alarms, indicators and monitoring equipment

The system electronics should include some indicators which display the state of the system, or at least its main parameters. The main indicators should display the low-charge state for the batteries and the overcharge.

In some instances, the user should be warned about the state of the system by an alarm. An example of this situation could be when the battery charge is low, and the charge limiter is about to cut off the supply.

The operation of many photovoltaic systems is monitored and the values of various parameters registered. For example, it is useful to have information about the energy collected and supplied, and about the irradiance, temperature and voltage produced by the system.

Summary

The power conditioning and control elements make it possible to convert the generated DC power to AC, protect the battery against overcharge or excessive discharge, and optimise the energy transfer between the PV generator and the battery or load. Indicators and monitoring equipment were also briefly discussed.

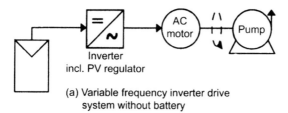

(a) Variable frequency inverter drive
    system without battery

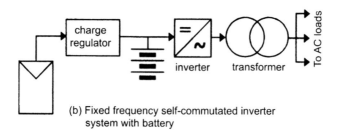

(b) Fixed frequency self-commutated inverter
    system with battery

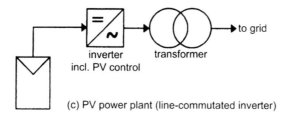

(c) PV power plant (line-commutated inverter)

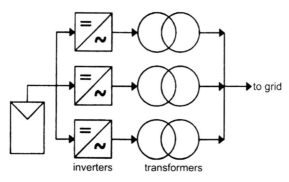

(d) Large multiple inverter PV power plant

Fig. 4.20 Various inverter configurations

4.6 SIZING PHOTOVOLTAIC SYSTEMS

4.6.1 Introduction

Sizing of a PV system, particularly a stand-alone one, is an important part of its design. Since the capital equipment cost is the major component of the price of solar electricity, oversizing the plant has a very detrimental effect on the price of the generated power. Undersizing a stand-alone system, on the other hand, reduces the supply reliability.

The sizing of a system requires a knowledge of the solar radiation data for the site, the load profile, and the importance of supply continuity. In addition, other constraints on the design (for example, economic) must also be known. The sizing procedure then recommends the size of the photovoltaic generator and battery capacity that will be optimum for the application. It will also allow the nominal characteristics of the electronic components to be specified.

4.6.2 Radiation and load data

Radiation data for a particular site (when available) are usually given in the form of global irradiation on a horizontal surface. We have shown in Chapter 2 how this information, in the form of monthly averaged daily data, can be translated into the radiation received by a panel at an inclined orientation (Table 2.2 and Figs 2.9 and 2.10). Some of these results are reproduced in Table 4.3.

Table 4.3 Daily irradiation in Barcelona
(data in kWh/m^2, panel inclination 60°)

January	4.60	July	4.80
February	4.59	August	4.85
March	4.62	September	4.58
April	4.55	October	4.53
May	4.30	November	4.12
June	4.45	December	3.49

Annual mean 4.46

The *load data* give detailed information about the appliances or equipment to be powered: their number, nominal power, nominal operating voltage and the number of hours of operation in a typical day.

To imagine what a typical load requirement might be, we give here some examples for home appliances. The appliances in question may be lights, washing machine, refrigerator, water pump, TV set, or other electric appliances. Every appliance can be described by its nominal power, voltage, and hours of operation in a typical house. Some of these data are summarised in Table 4.4.

Table 4.4 Data for typical loads

Appliance	Nominal power (W)	Nominal voltage (V)
Light appliance 1	15	24
Light appliance 2	20	24
Washing machine	100/800	220 rms
Refrigerator	70/150	220 rms
Small electric appliances	250	220 rms
Water pumping	100	24
TV set	60/100	24
Others (radio emitter/receiver)	120	24

Table 4.5 Typical loads for housing

System type	Energy supplied (Wh/day)	Charge at 24 V (Ah/month)
A	500	625
B	1000	1250
C	1500	1875

The size of the PV installation depends on the number and hours of operation of each of these appliances. One can distinguish various categories according to the energy demand per day as shown in Table 4.5.

Type A is a simple system providing energy to power five light appliances (at
an average of 1.65 h per day), a TV set in operation for 4 h per day, and the use
of small electrical appliances.

Type B is a small but sufficient installation to power 12 light appliances, a
DC washing machine (2 kg) with cold water, small electrical appliances, TV and
water pumping.

Type C is a size more adequate for housing and small agricultural needs.
More energy is therefore allocated for water pumping and more light appliances. A washing machine (220 V AC) can also be powered.

In addition to the average power demand of the load, the *load profile*—both
annual and daily—also has an important effect on the design of the system.
Clearly, a system powering a load whose peak matches periods of high solar
radiation will have smaller storage requirements. In some systems, the load can
be adjusted to respond to the energy input to the system. An example can be a
stand-alone domestic system where a washing machine is allowed to be used
only when there is enough sun.

4.6.3 The system energy balance

A relatively simple sizing procedure can be adopted for some, usually attended,
systems where supply reliability is not of paramount importance. The system

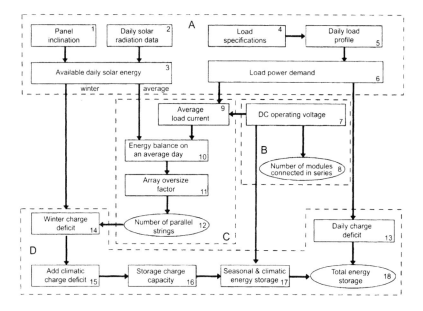

Fig. 4.21 Sizing procedure based on energy balance

design can then be based on the yearly energy balance between the radiation and the load. This is usually appropriate for locations where the variation of solar radiation during the year is not too pronounced. In regions with large seasonal variability of solar radiation input or for high reliability applications, the sizing procedure should be based on the month with the lowest irradiance (see Chapter 2).

We shall show how this sizing method is implemented for an array with a fixed panel orientation (Fig. 4.21) for a system without maximum power point tracker. The panel inclination can be chosen according to the general rules which were discussed in Chapter 2, or it can be treated as a parameter subject to optimisation in the final design.

A. Input to the sizing procedure

1–3. Determination of the energy input. The radiation data for the site, together with the panel orientation, are used to determine the incident solar radiation on the panel for a typical day in every month of the year (see section 2.3, unit 2).

4–6. Determination of the load demand. The load specification, or the typical load for a similar system, is used to determine the load power demand on a typical day. The load profile should be determined by estimating the times when various appliances will be needed. The energy which will be stored regularly on a daily basis to satisfy the load will make up the daily charge deficit which has to be allowed for in sizing of the storage subsystem (see [13]).

To make allowance for the battery efficiency being lower than unity, the load should be multiplied by

$$\frac{\eta_{bat}}{1 - f(1 - \eta_{bat})}$$

where η_{bat} is the energy efficiency of the battery, and f is the fraction of daily load which must be stored in the battery before being used.

B. Number of series-connected modules

7. The DC operating bus bar voltage V_{DC} of the system is specified. It is usual to take V_{DC} to be a multiple of the nominal battery voltage of 12 V.

8. As we have seen in section 3, the number of modules N_s which are to be connected in a series string is directly determined by the DC operating voltage. We can immediately write down

$$N_s = \frac{V_{DC}}{V_m} \qquad (4.7)$$

where V_m, the operating voltage of one module, should be taken as 12V for a module of 36 cells.

C. The number N_p of parallel strings

This number is directly related to the current requirement of the load.

9. The equivalent load current is calculated from the equation

$$I_L(A) = \frac{E_L}{24 V_{DC}} \qquad (4.8)$$

where E_L (Wh/day) is the typical power requirement of the load.

10. We now define the nominal current I_{PV} which is required from the photovoltaic generator when irradiated by the standard AM 1.5 radiation at $1\,kW/m^2$. The energy balance for a typical day can then be written as:

$$E_L(Wh/day) = PSH\ I_{PV} V_{DC} \qquad (4.9)$$

where we have expressed the radiation incident on the panel in *peak solar hours*, equal to the number of hours of the standard irradiance $(1\,kW/m^2)$ which would produce the same irradiation. Clearly, PSH is numerically equal to the irradiation in kWh/m^2 day. Using equation (4.8), equation (4.9) becomes

$$I_{PV} = \frac{24 I_L}{PSH} \qquad (4.10)$$

Equation (4.10) has a simple interpretation. The average load current, multiplied by the number of hours in the day, must be equal to the nominal current from the PV generator, multiplied by the number of peak solar hours. We have seen in section 4.3 that, in systems without maximum power point tracking, the nominal current is equal to the short-circuit current.

11, 12. The number of modules to be connected in parallel is then calculated using the following equation:

$$N_p = (SF)\frac{I_{PV}}{I_{SC}} \tag{4.11}$$

where I_{SC} is the short-circuit current supplied by an individual photovoltaic module when illuminated under standard conditions and (SF) is a *sizing factor* which is introduced to oversize the amount of current available from the array. We shall see below how the value of SF affects the overall design of the system.

D. Sizing of the storage subsystem

13. The daily and seasonal charge deficits are now calculated. It must be ensured that the night periods or shadows are covered satisfactorily. At the same time, excess energy generated and not used during the sunshine periods must be stored. This analysis determines the daily charge/discharge percentage of the battery that usually cannot exceed a given value for safe operation.

14. The energy balance for the year is set in such a way that the summer energy excess can be stored effectively to cover the cumulative energy deficit during the winter period. Thus, one calculates the winter energy deficit ΔE due to the seasonal weather cycle. This can be expressed as a charge deficit Q_{yd}, usually given in ampere-hours,

$$Q_{yd} = \frac{\Delta E}{V_{DC}} \tag{4.12}$$

This result is shown in Fig. 4.22 where the value of ΔE is represented by the shaded area. Note that this energy deficit depends on the choice of the array sizing factor SF.

15. A further climatic charge deficit Q is now added to allow for a number of days of operation without energy input (because of lack of sunshine, system maintenance, or servicing). This number – which characterises the number of successive days with below-average solar radiation – is determined from experience, and depends on the particular site of the system.

16. The total charge deficit due to the seasonal and climatic cycles is then determined, together with the application of pertinent safety factors:

$$Q_B = (Q_{yd} + Q_{los})\,(1/\Phi)(Ah) \tag{4.13}$$

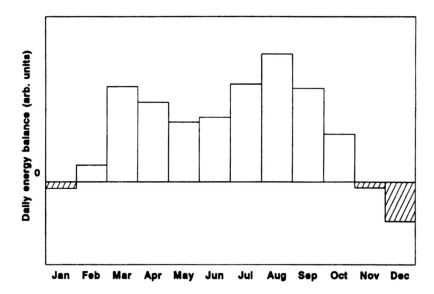

Fig. 4.22 The daily energy balance of a stand-alone PV system

where Φ is a factor that accounts for the fact that the batteries cannot be completely discharged. As we have seen in section 4.4, the value of this factor has an important effect on the battery life.

17. Given the DC operating voltage one can now determine the size of the energy store to cover the seasonal and climatic energy deficits.

18. The daily charge deficit is now added to the storage (with allowance for the safety factor Φ, see 16) and *using* manufacturer's specifications, one can determine the required size of the storage subsystem. This completes the preliminary system design.

One may, at this point, examine the free variables that can be optimised. For example, the effect of the panel sizing factor SF on the battery storage capacity should be examined (Fig. 4.23). Using this graph, one can choose the optimum array size and storage to minimise the cost of the system. Similarly, one can also optimise the panel orientation.

Once the storage capacity and panel size have been selected, it is recommended to simulate numerically the operation of the system using measured solar energy data, whenever available, and to determine the reliability of the power supply. An example of such procedure is shown in Fig. 4.24 which is applicable in situations where the energy is used in the evening or at night.

The model in Fig. 4.24 uses two constants: *storage* – the maximum amount of energy that can be stored in the battery and subsequently retrieved, and *demand* – the daily load. For each day, the amount of solar energy generated (*solar*) and the energy available in the battery (*stored*) form the input to the

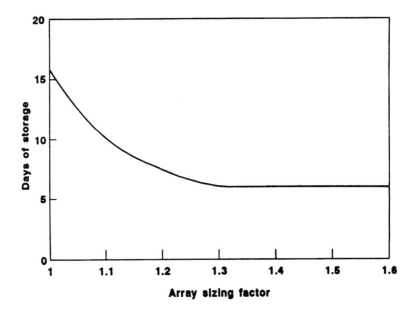

Fig. 4.23 The storage-array relationship based on energy balance

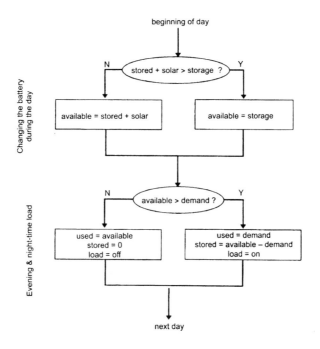

Fig. 4.24 A simple model of the PV system applicable for night-time load when long-term data are available. The charge regulator is represented by a simple device which disconnects the load when the stored energy drops below certain level (stored = 0 in the model)

routine. The output is the energy supplied to the load (*used*), the amount of energy available for the next day (new value of *stored*), and the state of the load (on or off) at the beginning of next day.

4.6.3.1 Worked example

In this example which illustrates the sizing method just described, the load current I_L has been set equal to 1 A. The calculation is then particularly transparent, and the design is quite general since the the current I_{pv} required from the PV panels scales linearly with I_L.

The calculation is carried out using the irradiation data for Barcelona which are given in Table 4.3. Let us suppose that we choose the panel inclination $\beta = 60°$. This is close to the value which maximises the energy input during the 'worst month' (here, December).

The average value for PSH from the data in Table 4.3 is 4.46 h and using equation (4.10) we obtain:

$$I_{pv} = 5.38 \text{ A} \qquad (4.14)$$

The number of parallel modules is calculated using equation (4.11),

$$N_p = \frac{5.38}{2.1} = 2.56 \qquad (4.15)$$

The nominal current of one module was assumed here equal to 2.1 A.

N_p is now set to 3. This gives an effective safety factor of SF = 1.17, and an effective oversizing of 17% in panel current.

Figure 4.25 shows the incident energy on the panels as a function of the month of the year. The mean value is 4.46 k Wh/m² day.

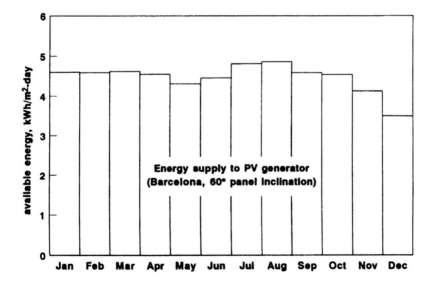

Fig. 4.25 Energy supply to photovoltaic generator

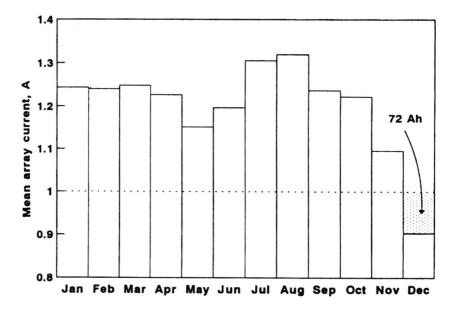

Fig. 4.26 The mean current produced by the photovoltaic generator over the year

Figure 4.26 shows the resulting current supplied by the three parallel panels. It is clear that the year mean value is not 1 A but 1.17 A as a result of the effective oversizing.

In the example, the winter deficit Q_{yd} is 72 Ah. Assuming a maximum period of 5 days without sunshine, $Q_{los} = 5 \times 24 \times 1\text{Ah} = 120$ Ah. A further charge capacity (a fraction of a day's load) could be added to compensate for an unfavourable daily load profile. Finally, assuming a conservative value of 70% maximum discharge, the capacity needed for the battery is set at 647 Ah. When choosing the battery one should remember that this capacity is under 100 h discharge rate.

4.6.4 Sizing and reliability

In many applications of photovoltaics today, reliability of supply is an important factor, and must be reflected in the sizing calculation. When reliability is important, there should be no seasonal charge deficit in the sizing procedure based on energy balance (section 4.6.3). In other words, the array sizing factor should be chosen sufficiently large so that the array can supply enough energy even during the 'worst month' (see section 2.2.3).

More sophisticated sizing procedures can give, as a final result, the cost effective combination of the array size and storage capacity, subject to a minimum acceptable level of supply reliability. The supply reliability is measured by the loss-of-load probability (LLP), defined as

Table 4.6 Recommended loss-of-load probability values

Application	LLP
Domestic:	
Illumination	10^{-2}
Domestic appliances	10^{-1}
Telecommunications	10^{-4}

$$LLP = \frac{\text{Average duration of supply interruption}}{\text{Total duration of supply}}.$$

Some examples of the LLP values for different applications are given in Table 4.6.

As we have seen in section 4.3, the design of a photovoltaic system can be verified to satisfy the necessary reliability requirements by calculating the LLP figure for the design. Methods are becoming available which allow one to input this information to the sizing procedure from the start, and generate a set of *isoreliability lines* (Fig. 4.27, Egido and Lorenzo, 1992). Each line is a result of sophisticated computer algorithm, and represents the combination of storage capacity and array area which satisfies the reliability level described by the LLP value. These curves can then be further analysed to find the optimum combination with respect to cost, and represent the ideal information needed for a system design. Since this type of analysis is, at present, available only for a limited number of sites, more intuitive methods are often used to install high-reliability PV systems in practice.

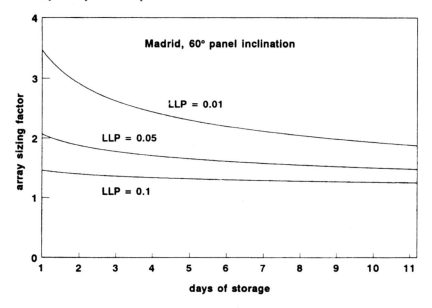

Fig. 4.27 Sizing based on loss-of-load probability

Summary

The sizing calculation determines the size of the array and battery for a specific application, and must take into account the following factors,

(a) power supply from the sun and the requirement from the load,
(b) maximum acceptable level of supply interruption, or loss-of-load probability (LLP), to allow for the random variation of supply and load,
(c) cost and other constraints relating to the system.

A sizing procedure based on the energy balance in the system was considered in detail, and a more rigorous method based on LLP outlined.

4.7 PHOTOVOLTAIC-DIESEL HYBRID ENERGY SYSTEMS

4.7.1 Introduction

In the past, electrical power has been generated in remote areas using engine-driven generators. For applications where a reliable, stationary generator set is required, diesel generators are generally preferred. For less frequent use, petrol generators may provide electricity at the lowest overall cost. Engine-driven generators are inherently inefficient when operated at light loads (below 40–50% of their rated capacity), which can also shorten their operating life and result in high maintenance costs. Low combustion temperatures during periods of light loads cause incomplete combustion and carbon deposits ('glazing') on the cylinder walls, leading to premature engine wear. Furthermore, conventional multiple-diesel remote area power supplies have to maintain a minimum level of *spinning reserve* (extra rotating capacity on line to overcome sudden load increases) while allowing sufficient time to bring another generator on line, which decreases their fuel efficiency.

More recently, the continuous decline of costs for renewable energy technology, together with the establishment of a mature alternative energy industry, has led to the increased utilisation of renewable energy sources for remote area power generation. Typically, photovoltaic modules and small to medium size wind generators are used, whereas small hydro-electric generators are only suitable in some locations. Combining renewable and conventional energy sources with a battery bank for storage, forming a hybrid energy system, can provide an economic and reliable supply of electricity. In comparison with a system which uses only a PV generator, larger systems are more common and the inclusion of a diesel generator reduces the size of the battery bank and improves the reliability of the overall power supply (reduces the loss-of-load probability discussed in section 4.6.4). Figure 4.28 shows a schematic diagram of a PV-diesel hybrid energy system.

Today, the most common application of hybrid energy systems is that of diesel generator augmentation, where the renewable energy source and the

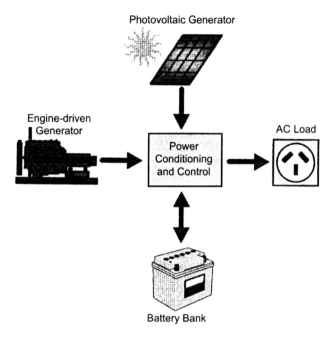

Photovoltaic Generator

Engine-driven
Generator

AC Load

Power
Conditioning
and Control

Battery Bank

Fig. 4.28 PV-diesel hybrid systems

battery bank are sized to reduce the run-time of the engine-driven generator. These systems provide sufficient storage to allow the load to be shifted, therefore ensuring that the generator is always substantially loaded. Generally, such systems are installed in locations where the logistics and economics of a reliable fuel supply are not a major contributing factor to the overall cost of system operation. The 'displacement-type' system is sized to reduce the fuel consumption of the diesel generator by 70–90% compared with a diesel-battery system, therefore relying heavily on the renewable resource such as solar. The engine-driven generator remains in the system to equalise the battery (see Section 4.4) and to act as a backup generator for extended periods of low solar input or high load demand. Such systems are generally installed in locations where fuel supplies are expensive and unreliable, or where other strong incentives for the use of renewable energy exist.

Hybrid energy systems have advantages over conventional, diesel-only remote area power supplies where the load demand over the day is highly variable (Fig. 4.29). A study of systems installed in the USA (Durand, 1996) concludes that hybrid energy systems are cost-competitive with conventional systems where the ratio of the peak load to the minimum load exceeds 3:1. If the load variability is less pronounced, then other constraints such as limited access or restricted environmental impact may favour the application of renewable energy sources for remote area power generation. While this general conclusion is not a substitute for a detailed life-cycle cost analysis

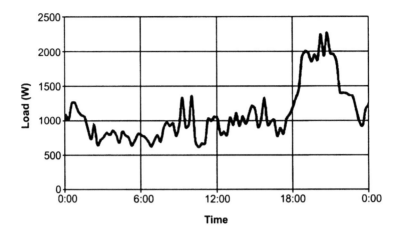

Fig. 4.29 Example of a daily load pattern for a remote area power supply (RAPS) system

of various system configurations for other locations or different operational conditions, it indicates that there is a large potential market for this emerging technology. It should be emphasised that the motivation for using this form of power generation comes not only from aiming for a stable electricity supply but also from the cost of supplying electricity in rural and remote areas.

Experience has shown that conventional diesel remote area power supplies are often not flexible enough to respond to changing load demand and varying operating conditions. In the past, this has resulted in compromises regarding the efficiency, reliability and supply quality, pending substantial upgrades of engine-driven generators. Significant changes of short-and long-term load demand are frequently experienced as a result of:

- increasing or decreasing population;
- changing consumer trends (e.g. increased use and number of electrical appliances);
- special community events;
- seasonal changes of environmental conditions (e.g. temperature, humidity).

In contrast, renewable energy sources and batteries are inherently modular and can be upgraded when the long-term load demand increases, without changing the overall configuration of the system. Power electronic converters, such as inverters, PV charge controllers, or battery chargers, should be sized such that an anticipated increase in load demand does not exceed their rated capacity. Alternatively, power conditioning devices can themselves be of modular design, which facilitates convenient system upgrades. System availability is improved by the inherent redundancy of multiple energy sources in hybrid energy systems. Either the inverter or the engine-driven generator may be used to meet the critical loads during unscheduled maintenance; this can be crucial for applications such as the refrigeration of vaccines or the supply of electricity for remote hospitals.

It needs to be emphasised that the generation of electricity in remote areas will remain expensive, regardless of the system type that is chosen. A life-cycle cost analysis of different supply options has to take into account the availability and cost of fuel over the entire lifetime of the system. It has to assess the renewable resources on a seasonal basis, consider the load demand, user preferences and requirements, and the availability of modern system technology together with adequate technical support.

4.7.2 System Configurations

Photovoltaic-diesel hybrid energy systems generate AC electricity by combining a photovoltaic array with an inverter, which can operate alternately or in parallel with a conventional engine-driven generator. They can be classified according to their configuration as (Nayar et al., 1993)

(a) series hybrid energy systems,
(b) switched hybrid energy systems,
(c) parallel hybrid energy systems.

An overview of the three most common system topologies is presented by Bower (Bower, 1993). In the following comparison, typical PV-diesel system configurations are described.

4.7.2.1 Series configuration

For the series system (Fig. 4.30(a)), the power generated by the diesel generator is first rectified and subsequently converted back to AC before being supplied to the load, which incurs significant conversion losses. During periods of low electricity demand the diesel generator is switched off and the load can be supplied from PV together with stored energy. AC power delivered to the load is converted from DC to regulated AC by an inverter or a motor generator unit. It should be noted that as a result of the series configuration most systems pass a large fraction of the generated energy through the battery, resulting in increased cycling of the battery bank and reduced system efficiency.

The actual load demand and the battery state-of-charge determine whether the diesel generator is operated. Depending on the power supplied by the photovoltaic array and the diesel generator, as well as the current load demand, the batteries are either charged or discharged. The solar controller prevents overcharging of the battery bank from the PV generator when the PV power exceeds the load demand and the batteries are fully charged. It may include maximum power point tracking to improve the utilisation of the available

(a)

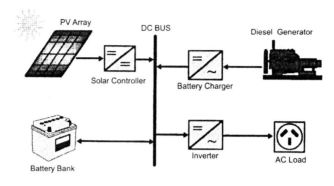

(b)

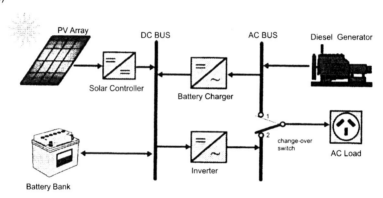

(c)

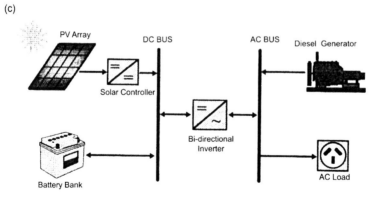

Fig. 4.30 Different configurations of PV-diesel hybrid energy system: (a) series configuration; (b) switched; (c) parallel

photovoltaic energy, although the energy gain is marginal for a well-sized system. The system can be operated in manual or automatic mode, with the addition of appropriate battery voltage sensing and start/stop control of the engine-driven generator.

Advantages

- The engine-driven generator can be sized to be optimally loaded while supplying the load and charging the battery bank, until a battery state-of-charge (SOC) of 70–80% is reached.
- No switching of AC power between the different energy sources is required, which simplifies the electrical output interface.
- The power supplied to the load is not interrupted when the diesel generator is started.
- The inverter can generate a sine-wave, modified square-wave, or square-wave, depending on the application.

Disadvantages

- The inverter cannot operate in parallel with the engine-driven generator. Therefore, the inverter must be sized to supply the peak load of the system.
- The battery bank is cycled frequently, which shortens its lifetime.
- The cycling profile requires a large battery bank to limit the depth-of-discharge.
- The overall system efficiency is low, since the diesel cannot supply power directly to the load.
- Inverter failure results in complete loss of power to the load, unless the load can be supplied directly from the diesel generator for emergency purposes.

4.7.2.2 Switched configuration

Despite its operational limitations, the switched configuration (Fig. 4.30(b)) remains one of the most common installations today. It allows operation with either the engine-driven generator or the inverter as the AC source, yet no parallel operation of the main generation sources is possible. Both the diesel generator and the PV array can charge the battery bank. The main advantage compared with the series system is that the load can be supplied directly by the engine-driven generator, which results in a higher overall conversion efficiency. Typically, the diesel generator power will exceed the load demand, with excess energy being used to recharge the battery bank. As for the series system, the diesel generator is switched off during periods of low electricity demand. Switched hybrid energy systems can be operated in manual mode, although the increased complexity of the system makes it highly desirable to include an automatic controller, which can be implemented with the addition of appropriate battery voltage sensing and start/stop control of the engine-driven generator.

Advantages

- The inverter can generate a sine-wave, modified square-wave, or square-wave, depending on the particular application.
- The diesel generator can supply the load directly, therefore improving the system efficiency and reducing the fuel consumption.

Disadvantages

- Power to the load is interrupted momentarily when the AC power sources are transferred.
- The engine-driven alternator and inverter are typically designed to supply the peak load, which reduces their efficiency at part load operation.

4.7.2.3 Parallel configuration

The parallel configuration (Fig. 4.30(c)) allows all energy sources to supply the load separately at low or medium load demand, as well as supplying peak loads from combined sources by synchronising the inverter with the alternator output waveform. The bi-directional inverter can charge the battery bank (rectifier operation) when excess energy is available from the engine-driven generator, as well as act as a dc–ac converter (inverter operation). The bi-directional inverter may provide *peak shaving* – the ability of parallel hybrid energy systems to supply loads that exceed the power rating of the engine-driven generator of the inverter from combined sources (see also section 5.7) – as part of the control strategy when the engine-driven generator is overloaded.

Parallel hybrid energy systems are characterised by two significant improvements over the series and switched system configuration.

(a) The inverter plus the diesel generator capacity rather than their individual component ratings limit the maximum load that can be supplied. Typically, this will lead to a doubling of the system capacity. The capability to synchronise the inverter with the diesel generator allows greater flexibility to optimise the operation of the system. Future systems should be sized with a reduced peak capacity of the diesel generator, which results in a higher fraction of directly used energy and hence higher system efficiencies.

(b) By using the same power electronic devices for both inverter and rectifier operation, the number of system components is minimised. Additionally, wiring and system installation costs are reduced through the integration of all power conditioning devices in one central power unit. This highly integrated system concept has cost advantages over a more modular approach to system design, but it may prevent convenient system upgrades when the load demand increases.

The parallel configuration offers a number of potential advantages over other system configurations. These objectives can only be met if the interactive operation of the individual components is controlled by an 'intelligent' hybrid energy management system where advanced system controllers may include Fuzzy Logic, Artificial Neural Networks, or Expert knowledge to optimise the decision making process.

Advantages
- The system load can be met in an optimal way.
- Diesel generator efficiency can be maximised.
- Diesel generator maintenance can be minimised.
- A reduction in the rated capacities of the diesel generator. battery bank, inverter, and renewable resources is feasible, while also meeting the peak load.

Disadvantages
- Automatic control is essential for the reliable operation of the system.
- The inverter has to be a true sine-wave inverter with the ability to synchronise with a secondary AC source.
- System operation is less transparent to the untrained user of the system.

4.7.3 System Components

The following section provides a brief overview of the main performance characteristics of the individual system components of photovoltaic-diesel hybrid energy systems. Components are discussed regarding their relevance to the control and operation of the system. The photovoltaic generator and battery storage were discussed in Sections 4.3 and 4.4. We therefore focus here on the remaining components of the hybrid systems, diesel generator and power conditioning.

4.7.3.1 Diesel generator

Diesel generators used in hybrid energy systems are synchronous alternators, which are directly coupled to a diesel engine. Adjusting the flow of fuel to the engine controls their operating speed, which determines the frequency of the AC output voltage. The operating cost of a diesel generator is governed by its fuel consumption, while the maintenance cost depends strongly on the number of operating hours and the loading of the engine. Frequent starting of diesel generators increases the mechanical wear on the engine. To minimise these costs the following requirements must be met by the diesel dispatch strategy.

(a) Once started the diesel engine should run for a minimum time period, typically exceeding at least 20 minutes of continuous operation. This performance specification is implemented to reduce engine wear and to minimise maintenance requirements.

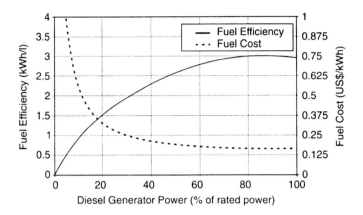

Fig. 4.31 Fuel efficiency and fuel cost of a diesel generator

(b) The diesel generator should not be allowed to operate below a minimum power level for an extended period of time. Typically, the minimum load is selected as 40% of its rated capacity, which prevents 'glazing' on the cylinder walls and avoids operation with low fuel efficiency.

A typical, medium size diesel generator has an optimum fuel efficiency of approximately 3 kWh/litre when the engine is run above 80% of its rated capacity. Operation of the diesel engine at lower output power reduces the fuel efficiency. This is shown in Fig. 4.31 which indicates the sharp increase in fuel cost below a minimum generator load of 40%. Inefficient operation of the diesel generator, which has to be sized to supply peak loads, is common for RAPS systems, due to the small base load experienced for many installations. A diesel fuel cost of US$0.5 per litre is assumed for the results shown in the graph. Depending on the remoteness of the location, the local infrastructure and fuel subsidies, as well as the global supply situation, the fuel cost may differ significantly from the calculated values.

In addition to the poor fuel efficiency of the diesel generator when operating at low load levels, the wear and corresponding maintenance requirements increase. The combustion chambers of the diesel engine do not reach their normal operating temperature at light loading, which results in carbon deposits on the cylinder walls and increased acidity of the lubricating oil.

4.7.3.2 Power conditioning

Three types of power conversion devices are typically included to condition and control the power flow in photovoltaic-diesel hybrid energy systems. In addition to the battery charge regulator and inverter (discussed in sections 4.5.3 and 4.5.5) a rectifier/battery charger is often included to convert AC power generated by the diesel generator to a regulated DC voltage which is used to recharge the battery bank. Series-type hybrid energy systems supply the AC load from the diesel generator via a two-step AC/DC and DC/AC conversion process. For example, assuming high efficiencies of 90% for both rectification and subsequent inversion of the DC voltage, this results in 19% of the generated power being lost due to the conversion process. This explains why switched and parallel hybrid energy systems achieve higher overall system efficiencies than series systems.

Typically, the diesel generator is operated at approximately 80% of its rated capacity to optimise its fuel efficiency. In switched or parallel hybrid energy systems AC loads are supplied directly from the diesel generator, with excess energy being used to recharge the battery bank. The electrical power that is supplied to the battery is controlled according to a defined battery charge strategy, which returns the battery to a high state-of-charge before the diesel generator is turned off and stand-alone operation of the system is resumed.

Modern parallel hybrid energy systems integrate the solar controller, inverter and rectifier in a single bi-directional inverter unit, which uses the same power electronic devices to implement both DC/AC and AC/DC conversion. Additionally, the automatic system management is implemented as part of the control functions of the micro-controller, which is required for the switching of the power electronic devices. The central system controller of a parallel hybrid energy system supervises the following tasks:

- automatic start/stop control of the diesel generator;
- continuous control of power flow;
- optimum loading of the diesel generator while it is operational;
- load-sharing between the inverter and the diesel generator at peak loads exceeding the rating of the diesel engine or the inverter;
- charge limiting of the PV generator when the batteries are at a high state-of-charge and the available PV power exceeds the load demand;
- temperature compensated, controlled battery charging from the diesel generator to ensure fast recharging while avoiding excessive 'gassing' of the batteries due to overcharge;
- controlled 'boost-charging' of flooded electrolyte lead-acid batteries at regular intervals (every 2–8 weeks) to reduce the negative effect of electrolyte stratification;
- disconnection of loads at low battery voltages to prevent excessive discharging.

4.7.4 Control and simulation

The design process of hybrid energy systems requires the selection of the most suitable combination of energy sources, power conditioning devices, and energy storage system together with the implementation of an efficient energy dispatch strategy. System simulation software is an essential tool to analyse and compare possible system combinations. The objective of the control strategy is to achieve optimal operational performance at the system level. Inefficient operation of the diesel generator and 'dumping' of excess energy is common for many remote area power supplies operating in the field. Component maintenance and replacement contributes significantly to the lifecycle cost of systems. These aspects of system operation are clearly related to the selected control strategy and have to be considered in the system design phase.

Advanced system control strategies seek to reduce the number of cycles and the depth-of-discharge for the battery bank, run the diesel generator in its most efficient operating range, maximise the utilisation of the renewable resource, and ensure high reliability of the system. Due to the varying nature of the load demand, the fluctuating power supplied by the photovoltaic generator, and the resulting variation of battery SOC, the hybrid energy system controller has to respond to continuously changing operating conditions. Figure 4.32 shows different operating modes for a PV single-diesel system using a typical diesel dispatch strategy.

Mode (I):   The base load, which is typically experienced at night-time and during the early morning hours, is supplied by energy stored in the batteries. Photovoltaic power is not available and the diesel generator is not started.

Mode (II): PV power is supplemented by stored energy to meet the medium load demand.

Mode (III): Excess energy is available from the PV generator, which is stored in the battery. The medium load demand is supplied from the PV generator.

Mode (IV): The diesel generator is started and operated at its nominal power to meet the high evening load. Excess energy available from the diesel generator is used to recharge the batteries.

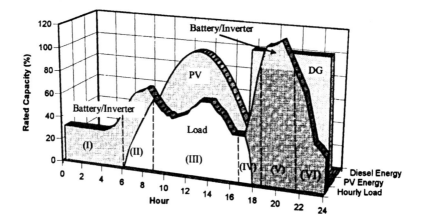

Fig. 4.32 Operating modes for a PV single-diesel hybrid energy system

Mode (V):   The diesel generator power is insufficient to meet the peak load demand. Additional power is supplied from the batteries by synchronising the inverter AC output voltage with the alternator waveform.

Mode (VI): The diesel generator power exceeds the load demand, but it is kept operational until the batteries are recharged to a high state-of-charge level.

In principle, most efficient operation is achieved if the generated power is supplied directly to the load from all energy sources, which also reduces cycling of the battery bank. However, since diesel generator operation at light loads is inherently inefficient, it is common practice to operate the engine-driven generator at its nominal power rating and to recharge the batteries from the excess energy. The selection of the most efficient control strategy depends on fuel, maintenance and component replacement cost, the system configuration, environmental conditions, as well as constraints imposed on the operation of the hybrid energy system.

Significant long-term performance improvements can be achieved through the application of optimised cycling procedures for lead-acid batteries. In most conventional applications, lead-acid batteries are charged either by a constant voltage or a constant current source. Various algorithms optimise the efficiency and duration of the charge cycle by implementing combinations of these two methods. High current charging (bulk charge) is performed until the battery reaches 75% to 80% of the maximum state-of-charge. Further charging with a high current is inefficient due to increased 'gassing' of the electrolyte, which is avoided by controlled charging with a reduced current. For hybrid energy systems this can result in inefficient operation of the diesel generator or reduced utilisation of the renewable resource.

The economics of hybrid systems can be estimated using the methods described in Section 5.2. The comparison of the life-cycle cost for different system configurations and operating strategies, however, requires detailed system simulations under realistic operating conditions. Packages which have been developed specifically for modelling hybrid energy systems, such as Hybrid2 from the National Renewable Energy Laboratory in the USA or RapSim from Murdoch University Energy Research Institute in Australia, and the reader is referred to the bibliography section for details about how these packages can be obtained.

Summary

Increasingly, hybrid energy systems are recognised as a viable alternative to grid-electricity or conventional, fossil fuel-based remote area power supplies. Rural households in industrialised and less developed countries attach a high value to a reliable, albeit limited, supply of electricity. Community facilities such as rural health centres, schools, or water pumping stations can contribute significantly to welfare and rural development. While it is recognised that technology can only be one aspect of community development, renewable energy systems have the demonstrated potential to provide some of the infrastructure needed in remove areas. Although the cost and technological development of hybrid energy systems in recent years has been encouraging, they remain an expensive source of electrical power. At present, PV-diesel hybrid energy systems are cost competitive for small to medium size systems where a reliable supply of diesel fuel is limited and expensive. Further improvements will allow the extension of markets for this emerging technology, both in industrialised and less developed countries.

Future challenges for the improved operation of hybrid energy systems are the reduced reliance on fossil fuel and the extension of the lifetime of the battery bank. Early replacement of batteries remains one of the most common problems experienced for many systems installed in the field, which leads to increased lifecycle cost and reduced availability of the system. Limited improvements in overall system performance can be achieved by further increasing the efficiency of power conversion devices, such as the inverter or the solar controller, mainly by reducing the losses at part load operation. Reliability is a difficult challenge for systems installed in a harsh environment, which is typical for many remote locations. Security of supply becomes a dominant design issue where technical support is limited and costly.

SUMMARY OF THE CHAPTER

The photovoltaic system usually consists of a number of subsystems. In addition to the photovoltaic generator, provision is usually made for energy storage and power conditioning, including control elements. Some systems also have a back-up generator.

The construction of a PV generator from modules has been discussed, focusing attention on the electrical characteristics of the generator under operating conditions and problems which arise from module interconnection.

The most common method of energy storage in PV systems is by lead-acid batteries. The characteristics of their operation have been analysed, alongside various chemical and physical processes which limit battery life or increase the maintenance requirements.

Power conditioning and control elements are often included to feed AC loads or the grid, interface different parts of the system, or monitor the system performance.

Sizing is an important part of the system design, particularly for standalone systems. The sizing procedure uses the radiation and load data to recommend the sizes of the array and battery storage, subject to the required reliability of power supply. It also allows the system cost to be minimised.

Photovoltaic/diesel energy systems which remain a popular option for remote area power supplies have been discussed, outlining system configurations, components and control.

BIBLIOGRAPHY AND REFERENCES

EGIDO, M. A. and LORENZO, E., The sizing of stand-alone PV systems: a review and a proposed new method, *Solar Energy Materials and Solar Cells* **26**, 1992: 51–69.
KALHAMMER, F. R., Energy-storage systems, *Scientific American* **241** (6), 1979: 42–51.
LORENZO, E. *et al., Electricidad Solar Fotovoltaica*, E.T.S.I. Telecomunicacion, Madrid, 1984.
MACOMBER, H. L. *et al., Engineering Design Handbook for Stand-Alone Photovoltaic Systems*, NASA contract DEN 3–195, 1981.

Further information about Hybrid 2 can be found at www.ecs.umass.edu/mie/labs/rerl

Details about RapSim can be obtained from Murdoch University Energy Research Institute, Murdoch University, South Street, Murdoch, Perth, Western Australia 6150. Tel. +61–8–9360 2868; Fax +61–8–9310 6094

SELF-ASSESSMENT QUESTIONS

PART A. True or false?

1. The hot-spot formation arises from non-uniformity of module encapsulant.

2. The blocking diode is connected in parallel with a string of modules.

3. Deep discharge improves the life of the lead-acid battery.

4. When modules are connected into a string, they should be selected according to similar short-circuit current.

5. The maximum power point of the module specified by the manufacturer determines the operating point of the battery.

6. The battery voltage increases when the load current is reduced.

7. The photocurrent depends exponentially on the temperature.

8. A DC/DC converter improves the interconnection between the AC/DC converter and the load.

9. Periodic overcharging of the battery improves the homogeneity of the electrolyte.

10. Once the LLP value is fixed, the relationship between the array size and storage size is described by a curve which resembles a hyperbola.

11. Diesel generators can most efficiently supply highly variable loads.

12. In a hybrid energy system, the battery can be used to increase loading on the diesel generator.

13. The fuel efficiency of a diesel generator decreases with the power supplied to the load.

14. Batteries in a hybrid energy system can eliminate the need for a maximum power point tracker.

15. The bi-directional inverter in a hybrid energy system converts AC power to DC power for battery charging.

16. While the diesel generator is operating in a parallel hybrid energy system the AC output from the inverter is synchronised with the diesel generator waveform.

PART B

1. What subsystems of a PV system can you name?

2. What are the main categories of PV systems?

3. How do the current and voltage of a module depend on the operating conditions?

4. How many cells usually comprise a module? How are the cells connected and why?

5. What is the role of a bypass diode?

6. What are mismatch losses and how can they be reduced?

7. What is the temporal pattern of battery operation in a PV system?

8. Why is it a good maintenance practice to overcharge the battery at the end of winter?

9. Is it true to say that the operational capacity of a battery in a PV system can be significantly higher than the manufacturer's specifications?

10. What are the disadvantages of operating battery storage as a seasonal buffer?

11. What is the role of the blocking diode? Is it always necessary?

12. How can a PV system operate without electronic control and what disadvantages does this system design have?

13. What types of charge regulators are used in PV systems?

14. Which inverter would you use in a stand-alone domestic system?

15. Which control element would you use to ensure maximum power extraction from PV generator?

16. Why is accurate sizing of PV installations important?

17. What data are needed to size a PV system?

18. What is a measure of reliability of electricity supply of a PV system?

19. Why is the panel inclination in a stand-alone system usually chosen steeper than in a grid-connected one?

20. Which parts of the block diagram in Fig. 4.21 are determined by the required reliability of the system? How does the design of the PV generator depend on the chosen DC bus bar voltage? How does it depend on the required current?

21. List the advantages and disadvantages of diesel based remote area power supplies.

22. Explain how a battery-diesel system can lead to more economic system operation compared with a diesel-only remote area power supply.

23. What are the possible operating modes of a bi-directional inverter in a parallel hybrid system?

24. Explain the concept of 'peak shaving' in a hybrid energy system.

25. Discuss how a dynamic energy tariff can lead to more efficient operation of a hybrid energy system.

26. Explain why a parallel hybrid energy system has a higher overall efficiency compared to a series system.

PART C

A rural dwelling at a location considered in SAQ of Part B, Chapter 2, is to be powered by solar electricity. Size a photovoltaic system to power the following electrical loads:

   2×20 W lights for 2 hours a day in summer, 4 hours in winter
   60W TV set to operate for 3 hours every evening.

(You may assume that 5 days of energy storage are needed for lack of sunshine, the battery is 85% efficient, and should only be discharged to 50% of its capacity.)

Answers

Part A

1, False; 2, False; 3, False; 4, True; 5, False; 6, True; 7, False; 8, False; 9, True; 10, True; 11, False; 12, True; 13, True; 14, True; 15, True; 16, True.

Part B

1. PV generator, storage, power conditioning, control and possibly a backup generator.
2. Stand-alone and grid-connected systems
3. The principal factors are voltage decrease with increasing temperature and current proportional to the irradiance.
4. Between 33 and 36 cells connected in series to charge a 12 V battery.
5. The bypass diode protects the modules from hot-spot formation and reduced output losses on partial shading or damage to the array.
6. Mismatch losses occur as a result of variable module or cell quality and because of partial shading of the array. They can be reduced by sorting the arrays prior to construction and by installing bypass diodes.

7. Cycling with daily and climatic cycles, possibly with a seasonal cycle.
8. To provide an equalising charge for the battery elements.
9. Yes. The battery capacity under slow discharge can be considerably higher than under the discharge rate specified by the manufacturer.
10. Long period of operation in low charge state may lead to electrolyte stratification and sulphation.
11. To prevent the array discharging the battery at night.
12. By setting the array operating voltage to fully charge the battery but in such a way that a further slight voltage increase produces a large drop in the charging current. The operation of such a system is unlikely to be completely satisfactory as is does not protect the battery against excessive discharge, and the power transfer from the generator will not be optimum at other temperature than that considered in the design.
13. Shunt (parallel) and series regulators.
14. Self-commuted inverter.
15. Maximum-power-point tracker.
16. Undersizing reduces the supply reliability and oversizing increases the cost.
17. Load and radiation data, and the required supply reliability.
18. Loss-of-load probability, LLP.
19. To maximise the energy collection in winter, thus reducing the storage requirements.
20. (a) Blocks 9 and 14. (b) The DC bus bar voltage determines the number of cells connected in series in each string. (c) It determines the number of parallel strings.
21. Advantages: well-established technology, reliable, straightforward system design. Disadvantages: high fuel and maintenance costs, noisy, air pollution, size must match peak load demand.
22. Operation of the diesel can now be managed so that it only runs under high load conditions.
23. As stand-alone inverter to supply the load, battery charging when diesel is lightly loaded, supplying the load when it exceeds the capacity of the diesel generator.
24. Load peaks are met by the battery in parallel with the diesel, hence the peak load seen by the diesel is 'shaved'.
25. Dynamic energy tariffs can encourage consumers to use energy when it is readily available from renewable resources by lowering the energy price at these times. This results in an increased direct use of the renewable energy and reduces the cycling of the battery bank.
26. The load can be supplied directly from the diesel generator, which avoids conversion losses due to AC–DC and subsequent DC–AC conversion.

Part C

Load $= 340$ Wh/day, increasing to 400 Wh/day with allowance for charging efficiency of the battery. Sizing using the worst month (November) data with **PSH** $= 3.88$ h yields the requirement of 103 W_p PV array (for example, 2×55 W_p modules). Storage of 5 days + 1 days for night-time load gives 4800 Wh battery $= 400$ Ah at 12 V.

5

Applications

AIMS

The aim of this unit is to introduce the most common applications for solar electricity.

OBJECTIVES

After completing this unit the student should be able to:

1. name a wide range of applications for photovoltaics,
2. explain the basic features of each application,
3. understand why photovoltaics is a suitable option in each case
4. assess the economic viability of a PV installation.

NOTATION AND UNITS

Symbol or acronym		SI unit	Other unit
A	Array area	m^2	
ALCC	Annualised life-cycle cost		
Ca	Annually recurring payment		
Cr	Single future payment		
d	Discount rate		
E	Daily subsystem efficiency		
E_m	Module efficiency		

Edited from manuscripts of O. Paish, B. McNellis and A. Derrick (Overview of PV applications, Economics, Water pumping), M. Hill and S. McCarthy (Domestic PV supply), B. Hill (Remote monitoring and control) and K. Bogus (Electric power generation in space). Additional information by A. Sorokin.

(Contd)

Symbol or acronym		SI unit	Other unit
E_s	System efficiency		
F	Array mismatch factor		
i	Excess inflation rate		
I	Yearly mean daily irradiation	J/m^2	Wh/m^2
N	Number of years		
Pa	Discount factor		
Pr	Discount factor		
PW	Present worth		

5.1 INTRODUCTION

There is an extensive range of applications where the PV system is already viewed as the best option for electricity supply (Fig. 5.1). These systems are usually stand-alone, and exploit the following advantages of PV electricity.

- There are no fuel costs or fuel supply problems.
- The equipment can usually operate unattended.
- It is very reliable and requires little maintenance.

The reason for PV success can be purely commercial, based on conventional economic arguments which will be discussed in section 5.2. Here belong the so-called professional applications (reviewed in section 5.5), and a wide range of consumer products such as electronic watches, calculators, power for leisure equipment and tourism, etc.

In satellite power systems (discussed in section 5.6) where price of the PV system is not the deciding factor, solar cells have been employed for some time by virtue of their superior technical merits, principally their reliability and low power/weight ratio.

An ever growing number of PV systems, however, are installed for their social benefit to rural communities in both the developing and industrial countries of the world. Various aspects of rural electrification will be discussed in section 5.3, with PV water pumping highlighted in section 5.4.

There has been much discussion in recent years about the possibility that photovoltaics will eventually be cheap enough to be economic for grid-connected applications, competing with conventional power stations. We shall discuss two possible scenarios, central and distributed generation, in section 5.7.

The evolution of the different applications is discussed in Box B5.1 'Photovoltaic Markets'.

RURAL ELECTRIFICATION

- lighting and power supplies for remote buildings (mosques, farms, schools, mountain refuge huts)
- power supplies for remote villages
- battery charging stations
- portable power for nomads

WATER PUMPING AND TREATMENT SYSTEMS

- pumping for drinking water
- pumping for irrigation
- de-watering and drainage
- ice production
- saltwater desalination systems
- water purification
- water circulation in fish farms

PV lighting for a Community Building

HEALTH CARE SYSTEMS

- lighting in rural clinics
- UHF transceivers between health centres
- vaccine refrigeration
- ice pack freezing for vaccine carriers
- sterilisers
- blood storage refrigerators

Floating Solar Powered Pump

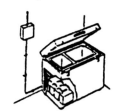

COMMUNICATIONS

- radio repeaters
- remote TV & radio receivers
- remote weather measuring
- mobile radios
- rural telephone kiosks
- data acquisition and transmission (river levels, seismographs)
- emergency telephones

PV Refrigerator for Vaccine Storage

AGRICULTURE

- livestock watering
- irrigation pumping
- electrical livestock fencing
- stock tank ice prevention

PV Power for a Radio-Telephone System

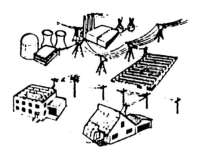

GRID-CONNECTED APPLICATIONS

- Distributed applications in buildings
- PV power stations

TRANSPORT AIDS

- road sign lighting
- railway crossings and signals
- hazard and warning lights
- navigation buoys
- fog horns
- runway lights
- terrain avoidance lights
- road markers

SECURITY SYSTEMS

- security lighting
- remote alarm system

PV Powered Navigation Buoy

CORROSION PROTECTION SYSTEMS

- cathodic protection for bridges
- pipeline protection
- well-head protection
- lock gate protection
- steel structure protection

MISCELLANEOUS

- ventilation systems
- camper and recreational vehicle power
- calculators
- automated feeding systems on fish farms
- solar water heater circulation pumps
- path lights
- yacht/boat power
- vehicle battery trickle chargers
- earthquake monitoring systems
- battery charging
- fountains
- emergency power for disaster relief
- aeration systems in stagnant lakes

PV Street Lamp

INCOME GENERATION

- battery charging stations
- TV and video pay stations
- village industry power
- refrigeration services

PV Powered Ventilator

ELECTRIC POWER FOR SATELLITES

- Telecommunications
- Earth observation
- Scientific missions
- Large space station

Fig. 5.1 Applications of photovoltaics

Box 5.1 Photovoltaic Markets

Production of solar cells and modules has been growing steadily throughout the 1990s, and the growth has accelerated towards the new millennium (Fig. B5.1). The most significant increases have been in the grid-connected applications, particularly photovoltaics in buildings. This market sector is gaining government assistance in a number of industrial countries, notably Japan, Germany, Switzerland, The Netherlands and USA. Several government programmes that drive this expansion – the Japanese 70 000 Roofs Programme, US One Million Solar Roofs Initiative and German 100 000 Roofs Programmes – are intended to provide the necessary volume for price reduction.

The other grid-connected sector – central power stations – is more sensitive to the installed system price. In the present climate of low oil and gas prices, substantial growth in this sector is not expected until considerable reduction of PV price.

There are 10 million small (less than 200 kW) diesel generators serving base loads throughout the sunbelt region of the world. Two million generators are sold every year. Photovoltaic systems are now economical for very small generators of less than 5 kW. Typical applications include water pumping, medical refrigeration, water purification and small village power.

Photovoltaic modules have proven to be the most reliable power source for remote unattended applications – particularly communications – throughout the world. This market is growing at a rate of 20–30% a year and is relatively inelastic to price. Crystalline silicon solar cells will continue to serve this important sector owing to its proven reliability. This sector is not dramatically affected by world energy prices or policy, and thus PV will continue to serve it as the most effective energy source.

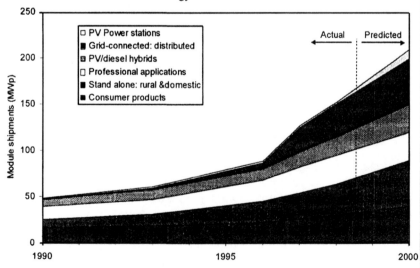

Fig. B5.1

Rural electrification is often seen as the greatest opportunity of all – electricity for the billions of people throughout the world without any electricity at all. However, there are major barriers to growth of this sector, as the persons with need have little or no money, and the cost of photovoltaic systems is capital intensive. In order to serve this vast market, in-country manufacturing is required to mitigate the balance of payments caused by the capital intensive photovoltaic production. India and China are the key examples of local PV module manufacture. Indeed, most developing countries impose import duties to protect local manufacturers. Over 50 000 small, 20–40W PV lighting and radio/TV systems are being installed each year in the developing world. The fastest growing application in this sector is the portable PV powered fluorescent lantern where over 110 000 were sold in 1996.

Consumer products represent one of the largest markets for solar cells. Most of these products – including calculators, watches, battery chargers, electronic remote controls and portable radios and lights – are powered by amorphous silicon solar cells, with Japan serving 90% of the market.
Adapted from Paul Maycock, PV News

5.2 ECONOMICS OF PV INSTALLATIONS

5.2.1 Introduction

The aim of this lesson is to learn how to calculate the economic value of a photovoltaic system so that it can be compared with other methods of power generation. The economic viability of a PV system must be assessed relative to the alternatives, for example, diesel, extending the grid, etc. At present, PV is most competitive where small amounts of energy are required far from the grid.

The economics of PV systems is different to that of other small power systems.

- The initial expenditure on the equipment (the capital cost) is high.
- There are no fuel costs.
- Maintenance costs are low.
- Reliability is high so replacement costs are low.
- The output of the system depends on its location.

The economic benefits of the system can either be money that has been saved, or revenue that has been collected, by operating the system. As well as economic benefits, there are also social and environmental benefits which should be taken into account when choosing between systems.

There are two ways of looking at the value of an electricity generating system. The economic approach takes the standpoint of the government, and so considers its value to the economy as a whole. It therefore looks at costs which exclude taxes and subsidies. In contrast, a financial assessment is an evaluation from the buyer's point of view. Therefore taxes, subsidies, interest payments on a loan, etc., must all be taken into account. Only economic assessment will be covered here, since that is the most general case.

In an economic evaluation, the following parameters are usually considered.

- *The life-cycle costs*. The sum of all the costs of the system over its lifetime, expressed in today's money.
- *Payback period*. The time it takes for the total costs to be 'paid for' by the monetary profits and other benefits of the system.
- *Rate of return*. The magnitude of the profits and benefits expressed as a percentage annual return on the initial investment.

Payback period and rate of return have two disadvantages. Firstly, it is not always easy to express the benefits gained in monetary terms, and secondly, they do not take account of how long the system will last, nor any future costs that will be incurred as time goes by.

Life-cycle costing is the most complete analysis and is the usual method for determining whether an application is economic. This section will only concentrate on life-cycle costing.

Box 5.2 Financing Renewables

There are a variety of options for accessing funds in order to start up or supplement a scheme for financing renewable energy projects. The source of funds has a critical impact upon how a fund is managed and how it is able to achieve self-sustainability. Funding for revolving loans and other financing schemes comes from three sources: bank funds, client savings and third-party financing. There is no right or wrong way to mobilise funding and provide services to clients, provided that certain criteria are met.

Developing countries share a number of characteristics in the development of their financial sector. Existing lending mechanisms vary from country to country, affected by regulations and the development of the financial sector. There have been many attempts to create new financing organisations in rural areas of developing countries, though many of these have not worked and these have not been for PV. There are important lessons to be learnt from the last 20 years of micro-enterprise and agricultural lending, and their failures and successes. This brings the conclusion that there are many different types of financing schemes that could be applicable to facilitate the purchase of PV systems. Whilst experiences show that there is no one scheme which will be applicable for all situations and countries, there are commonalties and indicators for choosing the 'best-fit' financing scheme.

Financing Options In Developing Countries

One of the major barriers to the deployment of privately-financed renewable energy systems in rural areas of developing countries is the initial purchase cost of these systems. We have seen in Section 5.2 that photovoltaic systems compare favourably with other rural electricity sources on the grounds of life-cycle costing. If appropriate financing were available, the barrier to wide PV system usage could therefore be largely overcome.

In most developing countries, there are no formal financial institutions in rural areas, i.e. the areas with the greatest need for small-scale renewable energy systems, such as PV systems for lighting and to power small industry. The reasons for this include:

- low population density,
- low $ loan requirement.

A low knowledge level about PV and other renewables in rural areas, and minimal or non-existent technical delivery mechanisms often compounds the problem caused by lack of appropriate financing systems. Often there is a perception that the electricity grid will come, and it will be free. In most cases, these two latter assumptions are false. If the grid does come, it is neither free nor less expensive than the electricity from renewables.

However, there are often lending organisations in place in rural and peri-urban areas, for which it would be appropriate to lend for such systems. Indeed, some are already doing so. These are usually in the semi-formal and informal financial sectors – though there are some examples of formal sector lending.

Informal sector
The generally-accepted definition of the informal financial sectors is that it is unregulated by government financial controls. This does not mean that it is not registered per se. Indeed, there are many financial organisations in the informal sector which have very strict rules of operation and reporting. For instance hire purchase, leasing companies and ESCOs are usually classified as being in the informal sector. However, there are usually strict corporate rules and regulations that they must adhere to in order to continue operating. Similarly, but usually in unwritten terms, Non-Governmental Organisations (NGOs) and local credit co-operatives operate by the rules of the local community which they service. Even the money lender is regulated by social norms: he charges high interest on very short-term loans, but this is known by borrowers. If either of the parties defaults on the terms of the verbal financial agreement, then the other expects regulation.

This regulation even if it is informal and unwritten, means that these financial organisations can and do play an important role in facilitating development at a local level. There are many different ways in which the informal sector can finance renewable energy installations. For instance, Singer Corporation of Sri Lanka, has financed PV systems, as has Kwazulu Finance & Investment Corporation in South Africa. There are a number of NGOs which are also active in this area, such as Enersol (Dominican Republic and Haiti), Solar Electric Light Fund (China, Sri Lanka, India), and Solanka (Sri Lanka). The Tuvalu Solar Electric Co-operative Society, and rural credit co-operatives in The Philippines, are examples of other organisations providing credit in the informal finance sector.

Leasing companies

The advantage to the end-user of leasing a PV system is that there is no down-payment. The lease payment usually covers maintenance and replacement (i.e. no extra outlays), and the end-user is not responsible for the technical upkeep of the system. Soluz has been operating in the Dominican Republic for over 5 years, leasing out Solar Home Systems.

A new form of PV leasing is the pre-payment card system which was initiated in 1999 by Shell Solar and Conlog in South Africa. This system is designed around a fee for service basis, and combines a charge controller, security system and pre-payment device into a single control unit. The pre-paid token/magnetic card activation device becomes inactive unless use is prepaid. This system is supported by local renewable energy supply compan-ies (established by Shell Solar), and the ability of the user to buy token cards at convenient and safe local outlets. The user does not purchase the system (thus overcoming the problem of upfront cash payment), and the system has a full technical backup.

Semi-formal sector

The demarcation between formal and semi-formal sector, and semi-formal sector and informal sector are often blurred and vary between countries. For instance, in savings credit co-operatives and credit unions are usually classi-fied into the semi-formal sector, but in some countries are classified with the formal sectors. The characteristics common to many organisations lending within the semi-formal sector are:

- peer group lending,
- savings and loans,
- history of international seed capital,
- can be linked to the formal sector,
- members receive share of profits,
- can be linked into other organisations,
- can be regulated,
- well-established in many countries and in rural areas,
- do not have stringent collateral requirements.

In some countries, village banks fall into the semi-formal category, along with Rotating Savings and Credit Organisations and credit co-operatives. There are a number of semi-formal organisations already lending for small PV purchases. These include BANPRES (Indonesia), Sarvoyda (Sri Lanka), Women's Union (Vietnam) and rural co-operatives in Bolivia.

Formal sector

There are very few examples of banks providing loans for small renewable energy systems. The reasons for this are that large financial institutions have high overheads. These organisations thus have high lending rates for small customers. Banks favour large-scale borrowers in urban areas, as these are more profitable than small borrowers in rural regions. Banks also require high collateral security on loans, which the poorer sections of the community cannot provide.

Commercial banks in developing countries which have financed PV projects include the Syndicate Bank in India, The People's Bank of Sri Lanka, Bank Rakyat in Indonesia, and Caisse Nacionale de Credit Agricole, Morocco. Development Banks lending for PV include Grameen Bank (Bangladesh), Banco Popular (Dominican Republic), and Banco del Nordeste do Brazil.

International Aid And Multilateral Development Initiatives

Until 1995, there were very few international or NGO projects in developing countries which provided funding for rural renewable energy projects. Indeed, there were very few such initiatives anywhere in the world. Since then, there has been a marked change in attitudes towards the sustainability of renewable energy technologies as primary energy generators, and their more positive effect on the environment than fossil fuel or nuclear technologies. Multilateral development organisations and Organs and Agencies of the United Nations now have a number of projects and programmes which include renewable energy technologies in an integrated manner. Three of these are described below. These and other initiatives are described on the UNDP, IFC, World Bank and UNEP websites.

PVMTI
The PV Market Transformation Initiative (PVMTI) is a joint initiative of the International Finance Corporation (IFC) and the Global Environment Facility (GEF). In 1997, the initial country research was undertaken, and the rationale and methodology for PVMTI developed. In 1998, IT Power and Impax Capital were appointed as the External Management Team for PVMTI. PVMTI will operate until 2008 in Kenya, India and Morocco, and will utilise $25 million from the Global Environment Facility for near-commercial investments or financing that will assist in market development and overcome barriers to PV dissemination. It is expected that, with leveraging, around $100 million in total will be invested.

In December 1998, 45 investment proposals were made. The first investment was made in October 1999, being $2 million for the development and expansion of energy stores in India.

SDC
The concept of the Solar Development Corporation (SDC) is to stimulate the private sector in developing countries in order to overcome the barriers. It is anticipated that this will be achieved both through market development activities and through financing activities. SDC itself will work with local financial institutions and distributors to develop tailored schemes aimed at accelerating electrification for rural families. The Business Plan was finalised in December 1997. Fund Managers were appointed in 1999, and capitalisation began later in that year. This initiative is jointly funded by the World

Bank, International Finance Corporation and US foundations including the Rockefeller Foundation.

FINESSE

The World Bank in, collaboration with The Netherlands Ministry of Development Co-operation (DGIS), US Department of Energy and UNDP/ EAP initiated the FINESSE strategy (Financing Energy Services for Small-Scale Energy Users) in 1989. The strategy recognises the need to ensure that the appropriate, commercial, institutional and financial mechanisms are in place in order to deliver renewable energy and energy efficiency services.

A key objective of the FINESSE strategy is to ensure financing mechanisms are in place to bring renewable energy and energy efficiency into the mainstream of lending by the Multilateral Financing Institutions, such as the African Development Bank.

The first FINESSE studies were initiated in Asia in 1989. In 1996, the Southern African Development Community Energy Sector Technical and Administrative Unit (SADC-TAU) in co-operation with the United Nations Development Programme Energy and Atmosphere Programme (UNDP/ EAP) initiated a programme to provide assistance to SADC countries to deliver technically feasible and economically viable renewable energy and energy efficiency services through the FINESSE model.

The principal activity of the SADC FINESSE programme was to conduct pre-investment investigations necessary to identify the renewable energy and energy efficiency projects (and appropriate financing schemes) and develop 'bankable' business plans for implementation. The SADC FINESSE project commenced in Lesotho, South Africa and Zimbabwe in 1996. In 1997, Angola, Malawi and Namibia joined the programme.

Business plans for bankable projects were developed with appropriate local stakeholders. The project also developed an Alternative Energy Fact-book for each country, an Institutional and Infrastructure Paper and a Project Assistance Handbook.

Since 1998, UNDP has been supporting FINESSE activities at the Development Bank of the Philippines, to elaborate a national renewable energy project pipeline.

Cost Of The Alternatives . . .

The macroeconomic costs of not electrifying the rural areas include the political factors of the disadvantaged rural population, the social costs of poor health and education, the migration from rural to urban areas, and the economic opportunity costs of lost production of food and other goods and services. None of these costs appear as items in the budget of any Government Department, but are borne by the whole of society.

Microeconomic comparison shows that in many circumstances it is cheaper to use decentralised generation by photovoltaics, wind or biomass and avoid the cost of traditional grid distribution. Sunlight is by far the most widespread renewable energy source, and electricity supply by photovoltaics is the least cost technique for large areas of the developing countries. In some areas however, wind, small hydro-plants or biomass may have lower costs than photovoltaics, so it is very important to study the wind, hydro, biomass and solar resource in each region in order to determine the optimum technique for rural electrification (Hill and Gregory, 1997).

(Written by J. Gregory)

References and Bibliography

R. Hill and J. Gregory *The Macroeconomic costs of NOT electrifying rural areas* 14th European Photovoltaic Solar Energy Conference, Barcelona, 1997.

IT Power Ltd., *Electricity from Sunlight*, Department for International Development, London, 1997.

J. A. Gregory *et al.*, *Financing Renewable Energy Projects: A guide for development workers*, IT Publications, London, 1998.

A. Cabraal, M. Cosgrove-Davies and L. Schaeffer, *Best Practices for Photovoltaic Household Electrification Programs, Lessons from Experiences in Selected Countries*, World Bank Technical Paper Number 324, The World Bank, Washington DC, 1996.

A. K. N. Reddy *et al.*, *Energy after Rio: Prospects and Challenges*, UNDP in collaboration with International Energy Initiative, Energy Initiative, Energy 21, and SEI, New York, 1997.

5.2.2 Life-cycle costing

In this method, not just the initial costs, but all future costs for the entire operational life of the PV system are considered. The period for the analysis must be the lifetime of the longest-lived system being compared.

For instance, a PV array costs more to buy than a diesel generator, but the modules should last over 20 years. The diesel generator might last 10 years, using a certain amount of fuel each year. So in this case the analysis period is 20 years. In addition to the capital cost, the cost of a replacement diesel after 10 years, plus 20 years' worth of fuel must also be included for the diesel option. In addition, the costs of maintenance and repair for the two systems over the whole 20 year cycle must be included. Depending on the exact figures, either the PV or the diesel system will work out cheaper overall.

To make a meaningful comparison, all future costs and benefits have to be discounted to their equivalent value in today's economy, called their 'present worth' or PW. To do this, each future cost is multiplied by a discount factor calculated from the discount rate. A discount rate of 10% per year would mean that in real terms it makes no difference to a farmer whether he has $100 now or $110 dollars in one year's time. Therefore a cost of $110 dollars one year from now has a 'present worth' of $100.

5.2.2.1 Parameters

The calculation of life-cycle costs requires values to be known for the following items.

- *Period of analysis.* The lifetime of the longest-lived system under comparison.

- *Excess inflation (i).* The rate of price increase of a component above (or below) general inflation (this is usually assumed to be zero).

- *Discount rate (d).* The rate (relative to general inflation) at which money would increase in value if invested (typically 8–12%).

- *Capital cost.* The total initial cost of buying and installing the system.

- *Operation and maintenance.* The amount spent each year in keeping the system operational.

- *Fuel costs.* The annual fuel bill.

- *Replacement costs.* The cost of replacing each component at the end of its lifetime.

5.2.2.2 Calculation of present worth

There are two types of calculation that are used in life-cycle costing when expressing a future cost or benefit as its present worth. The first is used to calculate the present worth of a single payment, say the replacement of a battery after five years. The second is used to calculate the total net present worth of a recurring cost, such as annual fuel or maintenance costs. This is the sum of many discounted single payments over the analysis period of N years. To avoid the need for complex equations, the relevant PW can be found by multiplying the actual cost by a factor that can be found in Tables 5.1 and 5.2. The formulae used to calculate the tables are:

$$\text{Pr} = \left(\frac{1+i}{1+d}\right)^N$$

$$\text{Pa} = \left(\frac{1+i}{1+d}\right)\left(\left|\frac{1+i}{1+d}\right|^N - 1\right) \Big/ \left(\frac{1+i}{1+d} - 1\right)$$

Single payment. For a single future cost Cr, payable in N years time, the present worth is given by

$$\text{PW} = \text{Cr} \times \text{Pr}$$

Example. It is estimated that a new pump will be required for a certain solar pumping system in 10 years time. We will assume that a new pumpset presently costs $1000, that pump prices do not change relative to general inflation, and that the discount rate is 10%. Using Table 5.1 with $i = 0$, $d = 0.1$ and $N = 10$

(the number of years hence that the payment is to be made), this gives a discount factor Pr of 0.39. Therefore the present worth of this future cost is

$$PW = \$1000 \times 0.39 = \$390$$

Annual payment. For a payment Ca occurring annually for a period of N years the present worth is:

$$PW = Ca \times Pa$$

Example. The fuel costs for a particular diesel generator are \$50 per year, and it might be assumed that diesel fuel prices will rise at 5% above inflation. Assuming a discount rate of 10% and a length of analysis of 20 years, Table 5.2 (with $i = 0.05$, $d = 0.1$ and $N = 20$) gives a cumulative discount factor Pa of 12.72. The present worth of the diesel fuel costs is therefore:

$$PW = \$50 \times 12.72 = \$636$$

Table 5.1 Selected values of present worth factor Pr for a cost in N years time

Discount Rate (d)	Inflation Rate (i)	Factor Pr for given number of years				
		5	10	15	20	30
0.00	0.00	1.00	1.00	1.00	1.00	1.00
	0.05	1.28	1.63	2.08	2.65	4.32
	0.10	1.61	2.59	4.18	6.73	17.45
	0.15	2.01	4.05	8.14	16.37	66.21
	0.20	2.49	6.19	15.41	38.34	237.38
0.05	0.00	0.78	0.61	0.48	0.38	0.23
	0.05	1.00	1.00	1.00	1.00	1.00
	0.10	1.26	1.59	2.01	2.54	4.04
	0.15	1.58	2.48	3.91	6.17	15.32
	0.20	1.95	3.80	7.41	14.45	54.92
0.10	0.00	0.62	0.39	0.24	0.15	0.06
	0.05	0.79	0.63	0.50	0.39	0.25
	0.10	1.00	1.00	1.00	1.00	1.00
	0.15	1.25	1.56	1.95	2.43	3.79
	0.20	1.55	2.39	3.69	5.70	13.60
0.15	0.00	0.50	0.25	0.12	0.06	0.02
	0.05	0.63	0.40	0.26	0.16	0.07
	0.10	0.80	0.64	0.51	0.41	0.26
	0.15	1.00	1.00	1.00	1.00	1.00
	0.20	1.24	1.53	1.89	2.34	3.59
0.20	0.00	0.40	0.16	0.06	0.03	0.00
	0.05	0.51	0.26	0.13	0.07	0.02
	0.10	0.65	0.42	0.27	0.18	0.07
	0.15	0.81	0.65	0.53	0.43	0.28
	0.20	1.00	1.00	1.00	1.00	1.00

Table 5.2 Selected values of present worth factors Pa for an annually recurring cost

Discount Rate (d)	Inflation Rate (i)	Factor Pa for given number of years				
		5	10	15	20	30
0.00	0.00	5.00	10.00	15.00	20.00	30.00
	0.05	5.80	13.21	22.66	34.72	69.76
	0.10	6.72	17.53	34.95	63.00	180.94
	0.15	7.75	23.35	54.72	117.81	499.96
	0.20	8.93	31.15	86.44	224.03	1418.26
0.05	0.00	4.33	7.72	10.38	12.46	15.37
	0.05	5.00	10.00	15.00	20.00	30.00
	0.10	5.76	13.03	22.21	33.78	66.82
	0.15	6.62	17.06	33.51	59.44	164.68
	0.20	7.60	22.41	51.29	107.59	431.39
0.10	0.00	3.79	6.14	7.61	8.51	9.43
	0.05	4.36	7.81	10.55	12.72	15.80
	0.10	5.00	10.00	15.00	20.00	30.00
	0.15	5.72	12.87	21.80	32.95	64.27
	0.20	6.54	16.65	32.26	56.38	151.24
0.15	0.00	3.35	5.02	5.85	6.26	6.57
	0.05	3.84	6.27	7.82	8.80	9.81
	0.10	4.38	7.90	10.71	12.96	16.20
	0.15	5.00	10.00	15.00	20.00	30.00
	0.20	5.69	12.73	21.44	32.22	62.04
0.20	0.00	2.99	4.19	4.68	4.87	4.98
	0.05	3.41	5.16	6.06	6.52	6.87
	0.10	3.88	6.39	8.02	9.07	10.19
	0.15	4.41	7.97	10.85	13.18	16.58
	0.20	5.00	10.00	15.00	20.00	30.00

5.2.2.3 Life-cycle cost (LCC)

For each payment to be made during the lifetime of the system, the present worth can therefore be determined using the discount factors Pr and Pa. The sum of all PW is the total life-cycle cost of the system.

The process could end at this point, but there are two further ways of expressing the life-cycle costs that are more meaningful and easily understood.

5.2.2.4 Annualised life-cycle cost (ALCC)

This is the total LCC expressed in terms of a cost per year. However, the LCC cannot simply be divided by the number of years in the analysis, as this takes no account of the change in value of money due to inflation and interest rates. The LCC must instead be divided by the factor Pa from Table 5.2, found using the chosen discount rate, inflation rate of zero, and a number of years equal to the analysis period. This is really the reverse process of discounting, and the result is expressed in $/year for each system.

Table 5.3 is a calculation sheet for performing a life-cycle costing, filled in for a sample PV system. Table 5.4 is an economic comparison between a PV and a diesel system using life-cycle costing.

Table 5.3 Life-cycle costing calculation sheet

System description: *250 W Stand-alone domestic PV supply*	

Parameters

Period of Analysis = *20 years*   Excess Inflation $i = 0$ Discount Rate $d = 10\%$

Capital Cost

Hardware:	$·········1900·········
Installation:	$········· 400·········
Total:	$ 2300

Operation and Maintenance:

Annual Cost	$········· 50········· per year
Discount Factor (Pa)	$·········8.51·········
Present Worth:	$ 425

Fuel

Annual Fuel Costs:	$········· Nil········· per year
Discount Factor (Pa):	····················
Present Worth:	$ Nil

Replacements

Item	Year	Cost	Pr	PW
Battery	5	$500	0.62	$310
Battery	10	$500	0.39	$195
Battery	15	$500	0.24	$120
			Total	$ 425

Total Life-cycle Cost	$ 3350
Annualisation Factor (Pa)	······8.51······
Annualised Life-cycle Costs	$ 394 per year

Table 5.4 Life-cycle costing comparison: PV vs diesel generating systems

Period of analysis	20.00 years
Discount Rate	10.00 %
Annualisation Factor	8.51

PV Calculation		Diesel Calculation	
Load	2.00 kWh/day	Using 5 KVA Low Speed Diesel Engine	
Battery Efficiency	70.00 %		
Demand at Array	2.86 kWh/day	Load	2.00 kWh/day
Days of Battery Storage	5.00 days		Lifetime
Design Insolation	5.50 kWh/m² day	Generator	3000.00 $ 20 yrs
Array Mismatch Factor	0.90	Installation	600.00 $
Module Price	4.50 $/W$_p$	Installed Capital Cost	3600.00 $
Battery Price	100.00 $/kWh		
Array Size	577 W$_p$	O&M Costs	360.00 $
Battery Size	10.00 kWh	Life-cycle O&M	3064.88 $
		Lifetime	
Array Cost	2597.40 $ 20 yrs	Diesel Price	0.26 $/litre
Battery Cost	1000.00 $ 5 yrs	Engine Efficiency	25.00 %
Support/Wiring	866 $ 20 yrs	Diesel Consumption	11.00 kWh/litre
Power Control	50.00 $ 10 yrs	Diesel Usage	0.73 litres/day
Capital Cost	4513.20 $	Diesel Cost	69.02 $/year
		Life-cycle Fuel Cost	587.59 $
Installation	902.64 $		
Total Installed Cost	5415.84 $	Replacement Costs	
O&M Costs	108.32 $/yr	Generator	0.00 $
Life-cycle O&M	922.16 $		
Recurring Costs		Total Life-cycle Cost	7252.47 $
Array	0.00 $	Annualised LCC	851.87 $
Battery	1245.86 $	Unit Electricity Cost	1.17 $
Support/Wiring	0.00 $		
Power Control	19.28 $		
Total Replacements	1265.13 $		
Life-cycle Cost	7603.14 $		
Annualised LCC	893.06 $		
Unit Electricity Cost	1.22 $/kW		

5.2.2.5 Unit electricity cost

Probably the most valuable figure for comparing two electricity-generating systems is the net cost of generating each kilowatt-hour during the lifetime of each system. This can be determined from the ALCC as follows:

$$\text{Electricity cost}(\$/\text{kWh}) = \frac{\text{ALCC}(\$/\text{year})}{\text{Electricity supplied (kWh/year)}}.$$

The electricity supplied each year can be estimated as:

$$\text{Electricity per year} = I \times AE_m \times E_s \times 365 \text{ kWh}$$

where I = average annual irradiation in kWh/m²-day

A = array area in m²

E_m = module efficiency

E_s = system efficiency

365 = number of days in a year.

When comparing two systems, it is often useful to see how the unit electricity cost varies depending on the size of the system, or to examine the effect of varying the module price, or the diesel fuel price. Graphs illustrating these variations can be computed by hand, or preferably using a spreadsheet program. An example comparing the electricity cost of PV with that of diesel or an extension of the grid is shown in Fig. 5.2.

The life-cycle costing process may appear rather long and complex, but if a little care is exercised it provides a relatively straightforward way to make a valid economic comparison between different options. Taken with other non-economic factors, this forms a vital part of the decision-making process.

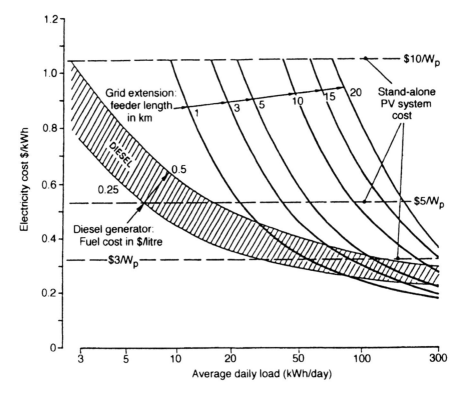

Fig. 5.2 Unit electricity cost against load. Comparison of PV, diesel and grid extension

5.2.3 Conclusion

The calculation laid out in Table 5.4 indicates that a PV system would be marginally more expensive than the diesel alternative. However, with the margin of error of a 20 year calculation, the two systems should be regarded as similar. It must be stressed that this calculation alone should not determine which system is the better option. There are other factors to be taken into account, in particular the following.

- *Reliability*. Although estimated costs for operation and maintenance (O&M) and component replacement were included in Table 5.4, further consideration must be given to the overall effect of a system breakdown. What will be the consequence of losing power for a number of hours or days? Depending on the application, this might mean lost revenue, decay of refrigerable goods, a lack of water, etc. In general, PV is considerably more reliable than the alternatives and this can greatly enhance its value to the customer.

- *Social and environmental benefits*. Other disadvantages of diesel systems which are hard to quantify but make them inferior to PV, are in areas such as noise, fumes, and vulnerability to fuel shortages. Against these one must also be aware of the greater land area required by the PV system which may have to be paid for, and its dependence on weather conditions.

The graph in Fig. 5.2, computed for an insolation level of $5.5\,kWh/m^2$ day indicates that on the basis of life-cycle costs, diesel will be cheaper than PV for anything above the smallest of loads (2–3 kWh/day) at the current PV system cost of around $10/W. However, with the anticipated fall towards $3/W over the next 20–30 years, PV will become economically attractive at much greater loads.

Summary

The method of life-cycle costing is the usual method for determining the economic viability of an application. We have shown how this method can be used to calculate the annualised life-cycle costs as well as the cost of generated electricity for PV systems, and thus provide an important input to the design of the system. An economic comparison has been made between PV systems and alternative energy sources.

5.3 RURAL ELECTRIFICATION

5.3.1 Introduction

The provision of electricity to rural areas derives important social and economic benefits to remote communities throughout the world. Power supply to remote houses or villages, electrification of the health care facilities, irrigation and water supply and treatment are just a few examples of such applications.

The potential for PV-powered rural applications is enormous. The UN estimates that two million villages within 20° of the equator have neither grid electricity nor easy access to fossil fuel. It is also estimated that 2 billion people worldwide do not have electricity, with a large number of these people living in climates ideally suited to PV applications. Even in Europe, thousands of houses in permanent occupation (and yet more holiday homes) do not have access to grid electricity.

As we have seen in section 5.2, the economics of PV systems compares favourably with the usual alternative forms of rural electricity supply, grid extension and diesel generators. The extension and subsequent maintenance of transmission lines over long distances, and often over difficult terrain, is expensive, particularly if the loads are relatively small. Regular fuel supply to diesel generators, on the other hand, often presents problems in rural areas, in addition to the maintenance of the generating equipment.

5.3.2 Domestic supply

Stand-alone PV domestic supply systems are commonly encountered in developing countries and remote locations in industrialised countries. The size range varies from $10\,W_p$ (for a single lamp) to $5\,kW$ depending on the existing standard of living. Typically larger systems are used in remote locations or island communities of developed countries where household appliances include refrigerator, washing machine, television and lighting. In developing regions large systems ($5\,kW$) are typically found for village supply while small systems (20–200 W) are used for lighting, radio and television in individual houses. In the developing world it is now estimated that there are at least 400 000 small Solar Home Systems which means that some 2 million people now rely on solar energy for their electricity.

A typical configuration of these systems is shown in Fig. 5.3. The inverter is omitted in systems where AC power is not required.

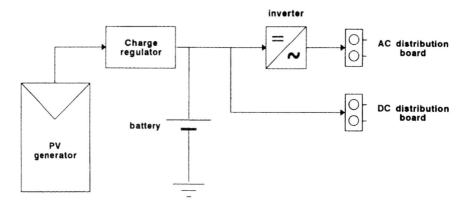

Fig. 5.3 Configuration of a stand-alone residential PV system

Example: Beginish Island Electrification (Fig. 5.4(a)). Jim and Mike Casey, two elderly brothers, are the only inhabitants of little Beginish Island off southern Ireland. Up until 1983, they rowed to the mainland every Sunday for Mass and to get the battery charged. Then the Caseys got a PV system to charge the battery which now powers the television, two 12 V, 15 W fluorescent lights and a VHF radio.

The complete PV system consists of four 19.2 W modules, two 12 V car batteries, a diode and a voltmeter. Operation and maintenance simply involve occasionally cleaning the array modules and adding distilled water to keep the battery plates covered. The Caseys control charging by manually adjusting the operating time of the television, lights and radio according to the amount of sunlight. They prevent deep discharge by turning off all the appliances when the voltmeter reading drops to 11 V.

This simple PV system has operated reliably and continuously since 1983.

Example: Rappenecker Hof, Germany (Fig. 5.4(b)). This 17th century farmhouse, now converted to an inn, has been powered by a PV system since 1987. When the sunlight is strong, the PV system supplies all the energy needs of the inn. A diesel generator, originally used to power the inn, backs up the PV system when the energy demand is high and when the weather is bad. The combined system is just as reliable as the utility grid. For example, in 1988 the total energy consumption at the inn was 2797 kWh. Of this, 77% was supplied by the PV system and the remainder by the diesel generator. (Recently, a 1 kW wind generator was installed to reduce the demand for diesel back-up.)

5.3.3 Health care systems

Extensive vaccination programmes are in progress throughout the developing world in the fight against common diseases. To be effective, these programmes must provide immunisation services to rural areas. All vaccines have to be kept within a strict temperature range throughout transportation and storage. The provision of refrigeration for this aim is known as the *vaccine cold chain* (Fig. 5.5).

Vaccine refrigerators are required to maintain vaccine between 0 °C and 8 °C at all times. In addition, a separate freezing compartment is usually needed to freeze ice packs which are used for transporting vaccines in cold boxes. Kerosene refrigerators are unreliable, and both bottled gas and diesel often suffer from fuel supply problems. Solar photovoltaic refrigerators (Figs 5.6 and 5.7 (a)) have shown that they are able to provide a more sustainable vaccine cold chain. In particular, they offer greater reliability, longer equipment lifetime and better temperature control. Over the past five years, over 2000 photovoltaic medical refrigerators have been installed in Africa.

There are, however, also some disadvantages. The main ones are that the repair work requires a skilled technician, the planning and installation of the

(a)

(b)

Fig. 5.4 (a) The Beginish Island PV system; (b) Rappenecker Hof

system takes longer, and that the local staff have to be carefully trained to use the system.

A solar photovoltaic refrigerator is likely to cost around $3000–$4000, more than a kerosene unit ($600–$800). The latter, however, will use 0.5 to 1.0 litres of fuel per day, require frequent maintenance, and have a shorter lifetime. In

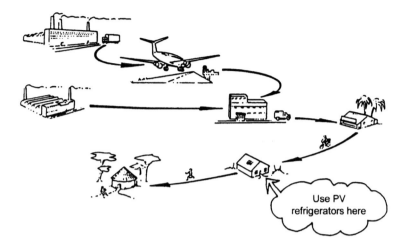

Fig. 5.5 The vaccine cold chain

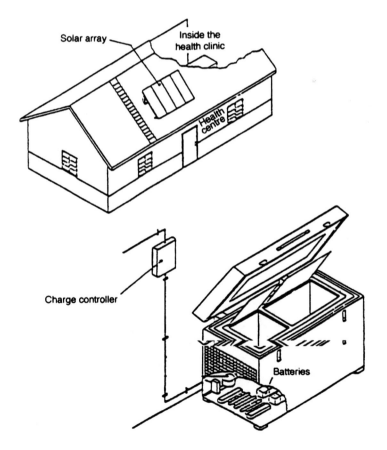

Fig. 5.6 Schematic diagram of PV refrigerator

general, the total cost over, say, 20 years will be approximately equal for the solar and kerosene refrigerators. But, because of the greater reliability and the resultant saving in wasted vaccine, solar refrigerators are the preferred option.

5.3.4 Lighting

Lighting is taken for granted in the industrial countries and in most of the urban areas of the developing countries. In areas without access to mains electricity, lighting is restricted to candles, kerosene lamps, or torches powered by expensive throw-away batteries.

In terms of the number of installations, lighting is presently the biggest application of photovoltaics, with tens of thousands of units installed world-wide. They are mainly used to provide lighting for domestic or community buildings, such as schools or health centres. PV is also being increasingly used for lighting streets and tunnels, and for security lighting (Figs 5.7(b) and 5.8).

The performance of PV lighting systems has been excellent, with increasing demand for more systems in the locality where a PV light is installed.

(a)   **(b)**

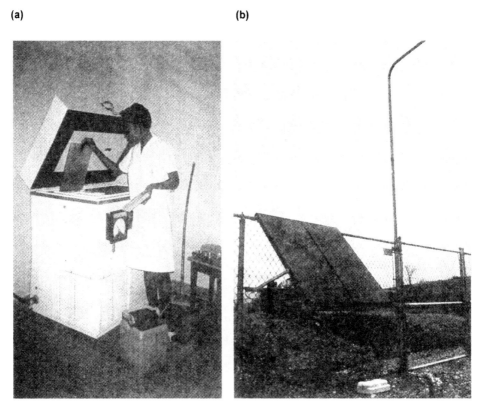

Fig. 5.7 (a) The vaccine refrigerator; (b) PV security lighting

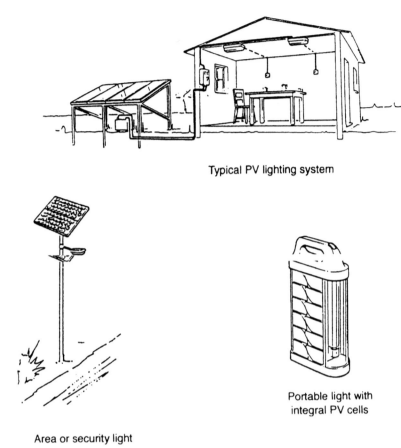

Typical PV lighting system

Portable light with
integral PV cells

Area or security light

Fig. 5.8 PV lighting systems

The use of low-voltage DC fluorescent lamps, rather than filament lamps, is important for efficient use of the electricity. AC lights can be used with an inverter but this introduces electrical losses.

Even though individual PV lights are more expensive to buy than kerosene lamps, they are generally more cost-effective over the lifetime of the installation. In addition, they provide better quality illumination.

5.3.5 Battery charging

In remote locations, batteries are often used as the main source of electricity. Ordinary car batteries are used not only in vehicles but also to power lights, televisions and radios. The usual method for charging batteries has been to use diesel or gasoline generators at battery-charging stations in the nearest village, or to transport the batteries periodically to a mains supply. PV systems are now

A 1400 W$_p$ battery charging station

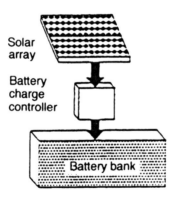

Fig. 5.9 PV battery charging

also common, either at battery-charging stations or in small dedicated systems (Fig. 5.9).

A PV battery-charging system consists simply of a PV array with support structure and charge controller. They are usually suitable for insolation levels above 3.0 kWh/m^2 day.

Summary

Photovoltaic systems provide a natural means for rural electrification. Various types of these systems have been described, with more detailed discussion devoted to domestic supply, health care systems, lighting and battery charging. Attention has been focused on the social benefits that this type of electrification brings to the rural communities.

5.4 WATER PUMPING

5.4.1 Introduction

More than 10000 PV-powered water pumps are known to be successfully operating throughout the world. Solar pumps are used to pump from boreholes, open wells, rivers and canals to provide water for villages and irrigation supplies. Less-common applications include drainage pumping and water circulation for fish farms. The largest number of solar pumps in one country is in India where more than 500 systems have been installed for village water supplies.

A solar-powered pumping system (Fig. 5.10) consists of a PV array powering an electrical motor which operates a pump. The water is pumped up through a pipe and into a storage tank. The energy from the PV array is therefore converted into the potential energy of the pumped water. In most instances, this obviates the need for battery storage of the generated electricity.

Solar pumps are used principally for two applications: village water supply (including livestock watering) and irrigation. These two applications have very different demand patterns: villages need a steady supply of water, whereas crops have variable water requirements during the year.

A solar pump for village water supply is shown schematically in Fig. 5.11. Note that water has to be stored for periods of low insolation. For example, in the Sahel region of Africa the storage needs to be 3–5 days of demand. During rainy seasons (which also coincides with the period of lowest solar radiation), the reduced output of the pump can be offset by capturing the rain water. The majority of solar pumping systems installed to date are for village water supply.

A solar pumping system for irrigation is shown in Fig. 5.12. In this application, the peak demand (during the irrigation season) is often more than twice the average demand. The solar pump is therefore under-utilised for most of the year.

5.4.2 The technology

Systems are broadly configured into five types (Fig. 5.13).

(a) *Submerged multistage centrifugal motor pumpsets* are probably the most common type of solar pump used for village water supply. The advantages of this configuration are that it is easy to install (often with lay-flat flexible

(a)

(b)

Fig. 5.10 (a) PV pump and storage tanks; (b) PV pumping with a concentrator array (courtesy of Midway Labs Inc.)

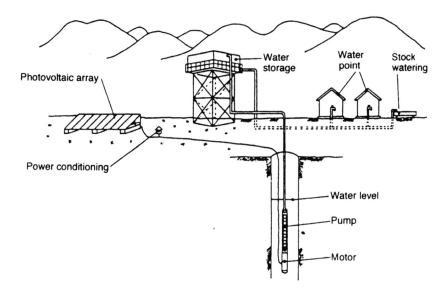

Fig. 5.11 Diagram of solar-powered village water supply

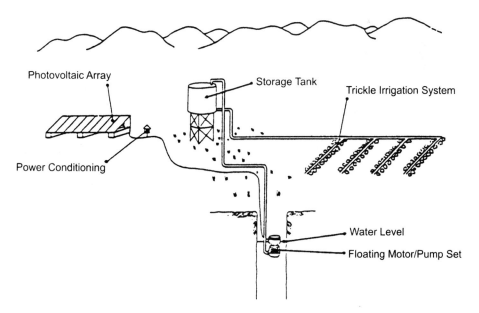

Fig. 5.12 Diagram of solar irrigation pumping system

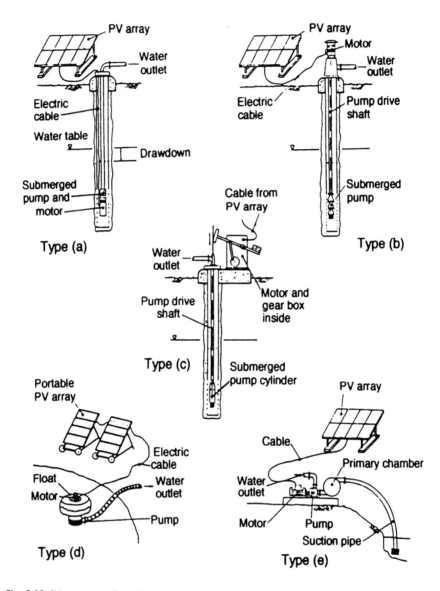

Fig. 5.13 PV pump configurations

pipework) and the motor pumpset is submerged away from potential damage. The most commonly employed system consists of an AC pump and inverter with a PV array of less than 1500 W, but DC motors are also used.

(b) *Submerged pumps with surface mounted motors* were widely installed with turbine pumps in the Sahelian West Africa during the 1970s. This configuration gives easy access to the motor for brush changing and other maintenance, but, in general, it is largely being replaced by the submersible motor and pumpset.

(c) *Reciprocating positive displacement pumps* (often known as the jack or nodding donkey) are very suitable for high-head, low-flow applications where they are often more efficient than centrifugal pumps. Reciprocating positive displacement pumps create a cyclic load on the motor which, for efficient operation, needs to be balanced. Hence the above-ground component is often heavy and robust, and power controllers for impedance matching are often used.

(d) *Floating motor-pump sets* have a versatility that makes them ideal for irrigation pumping from canals and open wells. The pumpset is easily portable and there is a negligible chance of the pump running dry. The solar array support often incorporates a handle or 'wheel barrow' type trolley to enable transportation.

(e) *Surface suction pumpsets* are not recommended except where an operator will always be in attendance. Self-start and priming problems are often experienced in practice, and it is impractical to have suction heads of more than 8 metres.

5.4.3 Sizing and cost

Sizing of solar pumps is determined by the hydraulic energy required:

Hydraulic energy (kWh/day)

$$= \text{Volume required } (\text{m}^3/\text{day}) \times \text{head } (\text{m}) \times \text{water density} \times \text{gravity}$$
$$= 0.002\ 725 \times \text{volume } (\text{m}^3/\text{day}) \times \text{head } (\text{m})$$

Solar array power required (kW_p)

$$= \frac{\text{Hydraulic energy required } (\text{kWh/day})}{\text{Average daily solar irradiation } (\text{kWh/m}^2\,\text{day}) \times F \times E}$$

where F is the array mismatch factor, equal to 0.85 on average, and E is the daily subsystem efficiency, typically between 0.25 and 0.4.

A PV system pumping $25\,\text{m}^3$/day through a 20 m head requires a solar array of approximately $800\,\text{W}_\text{p}$ in the Sahel region of Africa. Such a pump would cost approximately $6000. Other example costs have already been considered in section 5.2. A range of prices is to be expected, since the total system comprises the cost of modules, pump, motor, pipework, wiring, control system, array support structure and packaging. Systems with larger array sizes generally have a lower cost per W. The cost of the motor pumpsets varies according to application and duties; a low-lift suction pump may cost less than $800 whereas a submersible borehole pumpset costs $1500 or more.

In general, photovoltaic pumps are economic compared to diesel pumps up to approximately $3\,\text{kW}_\text{p}$ for village water supply and to around $1\,\text{kW}_\text{p}$ for irrigation.

Summary

Solar pumping is an important application of photovoltaics which does not need battery storage. Two important applications have been discussed, village water supply and irrigation. Different types of pumps have been reviewed, alongside their suitability for different uses. The sizing and typical prices of systems have been considered.

5.5 PROFESSIONAL APPLICATIONS

5.5.1 Introduction

The installations of PV systems for professional applications as discussed in this section are fully justifiable on the grounds of conventional economic arguments which were discussed in section 5.2. These applications include telecommunication systems (radio transceivers, telephone systems, etc.) (Fig. 5.14), remote monitoring and control (scientific research, seismic recording, climate monitoring, traffic data collection, etc.), signalling systems for railways

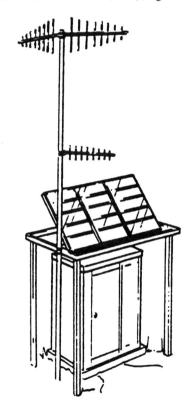

Fig. 5.14 PV-powered radio telephone system

or river and ocean navigation aids, cathodic protection of remote pipelines or bridges, and many other applications. Some selected applications of this type will now be discussed in more detail.

5.5.2 Telecommunications and remote monitoring

Photovoltaic systems have had more commercial success in telecommunications and remote monitoring than in other remote power applications. They can be effective at relatively low insolation levels compared to other applications. Figure 5.15 shows two examples of such systems — a PV-powered weather monitoring system and a hilltop TV translator.

Data collection from remote monitoring stations usually takes place by UHV/VHF radio transmission, when the central computer contacts the remote system for data. As well as transmitting information (for example, gas pressure) the system can also transmit the state of charge of the battery, thereby giving the central computer operator an advanced warning of system failure due to lack of battery power.

These applications need small but reliable power supplies which can be easily provided by a PV system charging a battery. Photovoltaics is ideal because the system can be sized to provide the small amounts of energy required for transmission and reception while requiring little or no maintenance.

The economics of PV are here very competitive with other forms of power supply. Mains power is unlikely to be available in the areas where remote monitoring systems are installed. Even if a mains cable were nearby, extending it more than a metre to supply a small amount of power is normally uneconomic.

(a)

(b)

Fig. 5.15 (a) PV-powered weather monitoring; (b) hilltop PV translator

Table 5.5 Example of array and storage sizes for remote monitoring applications in Nothern Europe, determined from the December Solar radiation data for an average year and for the worst year in ten (values given produce 1 Wh/day)

	Average	Worst
Array size (W)	13.5	110
Battery size (Wh)	50	165

Primary batteries are still the most commonly seen power source for remote systems. Initial cost comparisons with PV look attractive only until the long-term costs such as regular battery replacement, and the related personnel and vehicle costs, are accounted for.

To operate unattended for long periods, various hazards must be avoided which include theft or vandalism of the system, damage by animals or falling branches, shading by growing vegetation, and snow covering.

Since the reliability of these systems is paramount, the sizing method used is based on radiation data for the worst month of the year (usually December in the northern hemisphere) rather than on the average daily irradiation over the year. In addition, to ensure system operation even over periods of lower solar radiation than predicted from the average yearly values, the sizing often uses data for the worst year in ten. An example of system sizes that have been used in practice are given in Table 5.5.

Example (British Gas plc, UK). The system shown in Fig. 5.16(a) powers a UHF/VHF transmitter via a 12 V sealed lead-acid battery. The sizing of this system has been carried out according to the data in Table 5.5.

(a)

(b)

(c)

Fig. 5.16 (a) PV-powered remote monitoring; (b) OAD Beacon; (c) Cathodic protection of a gas well

Example (Guide Dogs for the Blind Association, UK). The GDBA orientation-assisting device consists of a series of beacons (as shown in Fig. 5.16(b)) strategically positioned along specific routes which transmit preselected location radio codes to a chosen distance. A receiver carried by a visually handicapped person responds to the beacon radio codes by playing a prerecorded audible message, giving orientation and directional information (where the shops are, road crossings, etc.).

Box 5.3 Solar Powered Marine Navigation Aids

Trinity House Lighthouse Service covers general aids to navigation around the coast of England, Wales, and the Channel Islands and, as a policy, adopted the use of solar photovoltaic power as an alternative to acetylene gas and diesel generation. This decision was taken in 1993 and was a difficult one bearing in mind the unreliable weather experienced generally throughout the country and possibly even worse in the coastal areas. To date, Trinity House have over 400 systems in the field and these vary from fixed sites (lighthouses) to floating aids to navigation covering buoys and lightvessels.

System Description

Fixed Installations
The fixed installations on mainland or island sites are designed around arrays that are facing due south and angled at latitude plus 15'. As we have seen in Chapter 2, this arrangement gives a system with maximum reliability throughout the year. There are, however, sites which, while being fixed, have insufficient suitable land mass to allow installation of the PV arrays in the optimum south facing mode. These sites are mainly 'rock towers' and the PV modules are fitted around the 360° perimeter of the tower (Figs B5.2 and B5.3).

Floating Sites
The floating installations are the most difficult to deal with as orientation and inclination are totally random, and an accurate performance prediction is

Fig. B5.2 Hanois Lighthouse (Channel Islands) automated 1996 and operating on solar power (courtesy of Trinity House Lighthouse Service)

Fig. B5.3 Inner Farne Lightouse (Northumberland) automated 1996 and operating on solar power (courtesy of Trinity House Lighthouse Service)

almost impossible. The floating aids can be divided into two types, light-vessels (ships) and buoys.

- **Lightvessels:** The solar arrays are arranged on each side of the vessel at the amidships position with $45 \times 38\,W$ PV modules per side. The arrangement is divided in such a way that half the modules from each of the power system (see below) are mounted within each array thus reducing the risk of poor performance (Fig. B5.4).

- **Buoys:** The solar modules are arranged around the superstructure, with the batteries mounted either in boxes on top of the buoy body or inside the superstructure itself, depending on the type of buoy. Battery protection against the elements is limited to preventing the full force of the waves swamping the battery. The theory used is water can get in but it can also easily get out (Fig. B5.5).

Fig. B5.4 Solar powered lightvessel (courtesy of Trinity House Lighthouse Service)

Electrical Systems
With the exception of buoys which have one 12 V power system, the other solar installations have a standard configuration of three separate power systems operating at 24 VDC centre.

- **Main System** which provides power to essential services such as aids to navigation and their controls. It has the largest capacity of the three systems with up to $2\frac{1}{4}$ kW of installed PV modules and 80 000 Wh of battery capacity.

- **Emergency System** is limited to providing power for emergency aids to navigation and is the smallest system with up to 150 W of PV modules and 8000 Wh batteries.

Fig. B5.5 Solar powered buoy (courtesy of Trinity House Lighthouse Service)

- **Facilities System** powers non-essential services covering station light-ing, miscellaneous alarms and communications. Typical installed capa-city here would be 700 W PV modules and 30 000 Wh battery.

Each of these systems is independent with no cross connections although on the load side certain equipment, for example, telemetry, has a second supply source in the event of its designated power supply failing. This ensures failure data is transmitted to the base control.

Design Parameters

In areas of high latitudes such as UK, it is important to ensure that the parameters in any solar PV system cater for the conditions in which the system is placed. Apart from latitude position, the UK climate is not conducive to a system that relies on regular and reliable solar irradiation. To regularise the design Trinity House have created a computer based Solar Model to ensure that systems meet the requirements.

- **Solar irradiation:** Fortunately, several UK meteorological stations have many years' data and are sited on or close to the coast. To cover area differences meteorological data are chosen from the 25 year rolling average statistics for the nearest station.

- **Energy demand** is a critical aspect and this is fed in to the programme using spreadsheet techniques with each load treated separately with duty cycles and operational hours fed in. It is interesting to point out that one of the heaviest loads, the fog signal, is the most difficult to predict for load profile,

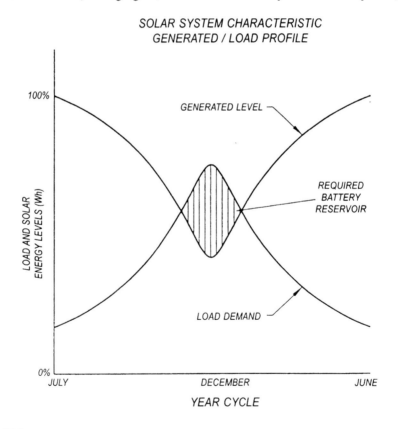

Fig. B5.6

The large difference between daylight hours in mid-summer and winter which, of course, is a function of latitude, also needs to be considered. This has a double effect of reducing the period of solar irradiance in the winter and, because a high proportion of the load is only activated at night, the winter load energy consumption is higher. This, of course, represents a deficit which is dealt with by the battery (Fig. B5.6).

Another effect of the climate is that on top of the deficit from the difference between winter and summer, there are periods of 'high pressure gloom' occurring in the winter when up to 30 days of zero or near to zero solar irradiance is experienced. Again this must be absorbed by the battery. To illustrate this the seasonal cycle is shown on Fig. B5.7. Here it is seen that there is a seasonal cycle over the year upon which is superimposed the daily cycle which, in our system, represents 1% to 2% of battery capacity.

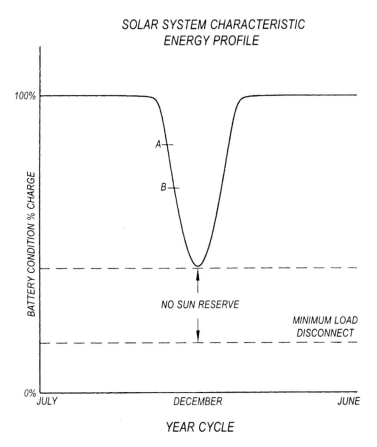

SOLAR SYSTEM CHARACTERISTIC
ENERGY PROFILE

YEAR CYCLE

ACTUAL PROFILE IS AS INDICATED, WITH
DAILY CYCLES OF CHARGE / DISCHARGE
OF APPROX. 1% TO 2% OF SYSTEM CAPACITY.

Fig. B5.7

From equipment parameters the computer sizing programme provides the answer to various 'what if' scenarios. For example:

- what if the allowance of 30 days without sun is reduced to 20?
- what if the modules are positioned at an inclined/vertical/horizontal orientation?

This allows the engineer to assess the particular site difficulties. For instance, a readily accessible site could accept less than 30 days no sun reserve and be more cost effective.

System Component

Solar Modules
It was considered that two distinct applications were involved, the land based and sea-borne. The land based, while being subjected to salt laden atmosphere, in general could utilise standard mono- or polycrystalline PV modules without any special precautions. With sea-borne units considerable attention was considered necessary with the mechanical aspects of the modules which would have to withstand shock, vibration and wave impact. Individual modules of a smaller surface area and with strengthened backing were chosen.

Batteries
As with the solar modules, a single battery type would not suit all applications, and three different scenarios have been identified.

- The buoy application is four or five years on station and subject to shock, vibration and sea water, together with large angles of inclination. The battery chosen was a valve regulated lead-acid type with gel electrolyte.
- The batteries on lightvessels also have to withstand shock and vibration, but have to have a life in excess of 12 years. Ventilation is no problem, low temperatures would not be experienced and after much research a traction type wet lead-acid battery was chosen.
- The land application is the least onerous. There are relatively stable temperatures, no shocks or vibrations, and no ventilation problems but the requirement is for possibly 15 years' life. The traction type battery could have been used but the attributes for shock and vibration were not necessary and costly. Tubular wet lead-acid type were chosen.

Charge regulators
The choice between using regulators or installing self-regulated systems was discussed in section 4.5. Trinity House decided to use regulators to protect the battery and to act as a load disconnect in the event of a discharged battery. However, an experimental system using the self-regulation technique (section 4.5.2) was also installed on a buoy and the results after four years of monitoring are encouraging. Regulators may therefore be dispensed with on future buoy installations.

The Load

Each item of load was analysed and methods investigated on reducing load energy consumption. One of the main moves was to use cycling for sensing devices. This does not detract from their performance but resulted in power reductions of 30% and above. 'Sleep modes' were also adopted again resulting in power savings.

CONCLUSION

Were Trinity House successful? The answer lies in the fact that over 400 individual solar PV systems have been installed in the field relying solely on solar PV power systems. It is not to say that there were not problems.

1. New fully charged batteries which, on test, were found to have less than the manufacturer's stated capacity. Batteries are now cycled and recharged immediately before going into operation.

2. Solar module frames out of alignment causing stress and eventual cracking of the modules.

3. During the exceptional weather conditions in October 1995, over 30 days of no sun were recorded, resulting in some marginal systems failing. All major stations are now fitted with emergency charging facilities.

4. The human element also came into play. Overkeen monitoring of performance over telemetry resulted in telemetry and communications energy consumption being far higher than anticipated.

WHAT OF THE FUTURE?

The use of solar PV systems will continue to expand within the Lighthouse Service, both as stand-alone and possibly as hybrid using other renewable sources as partners. This has been possible due to the advances in low power consumption electronics and higher efficiency light sources.

No move has been made to investigate solar systems on mains powered stations (grid connected) as the effort has been concentrated on conversion of gas powered and diesel generation stations. It remains to be seen whether the capital investment necessary for the solar PV installation on a mains station becomes cost effective.

The one negative effect of solar PV systems is the limited power availability and as a result the deterioration of the building fabric and internal fittings. The solution is to investigate the addition of alternative sources such as wind generation or the endothermic heating system – solar water heating with a heat pump. This, however, is in the early stages of implementation and its suitability is as yet unknown.

(Contributed by M. Wannell)

5.5.3 Cathodic protection

Cathodic protection is a method for shielding metalwork from corrosion, for example, pipelines and other metal structures. A PV system is well suited to this application for two reasons.

- Power is often required in remote locations along the path of a pipeline.
- Cathodic protection requires a DC source of electricity.

An example of such a system in shown in Fig. 5.16(c).

Corrosion is a chemical reaction between metal, usually iron, and water, which turns the iron into iron oxide (i.e. rust). You will remember from your chemistry lessons that oxidation involves electron transfer to the metal. If an electricity source is connected to the metalwork which sets it at a negative voltage (thus making it a 'cathode'), the iron is prevented from oxidising. The positive terminal of the source must be connected to an anode, often a piece of scrap metal, which will corrode instead.

Even if cathodic protection is used, the pipeline or structure must still be protected in the normal way with weather-resistant paint. Otherwise the PV array would have to be very large to cope with the high rate of corrosion.

Summary

A range of commercially viable applications of PV systems has been discussed. These systems usually operate unattended and often in isolated locations. The need for reliability in sizing has been emphasised.

5.6 ELECTRIC POWER GENERATION IN SPACE

5.6.1 Introduction

Photovoltaic solar generators have been and will remain the best choice for providing electrical power to satellites in an orbit around the Earth. Indeed, the use of solar cells on the US satellite Vanguard 1 in 1958 demonstrated beyond doubt the first practical application of photovoltaics. Since then, the satellite power requirements have evolved from a few watts to several kilowatts (Figs 5.17 and 5.18). After years of moderate growth of the space PV market the evolution of large-scale applications has accelerated significantly in the late nineties. The main market segment is dominated currently by telecommunication satellites with individual power levels above 10 kW. An increase to 20 kW and more is already under development.

Another large market segment is dominated by constellation programmes which provide multimedia, navigation and other electronic services. Programmes like Iridium, Globalstar and in future Skybridge, Teledesic and others make use of large numbers of satellites in orbit (ranging in number from 10 to several 100), representing accumulated solar array power of several hundred kW.

A unique high-power application is the Space Station Alpha which is being integrated in a low-earth orbit in the next few years (Fig. 5.19). Eight large-area solar arrays of 20 kW each will finally power Alpha.

A space solar array must meet strict criteria, and the most important considerations can be summarised as follows:

- extremely high reliability in the adverse conditions of the space environment,
- high power/weight ratio,
- cost constraints on the PV system are of secondary importance.

The satellite applications which are driving the European space photovoltaic technology include. Rosetta, a scientific deep-space mission presently under development (Fig. 5.20). On its flight from 2004 till 2014 it will visit the comet Wirtanen and travel at a distance from the sun more than 5 times the Sun–Earth distance. The more than 50 000 solar cells of Rosetta have to provide power at only 3% of one solar constant, at temperatures down to $-150\,°C$. This will be the first time worldwide that solar cells power a satellite under such difficult conditions in deep space. Special silicon and GaAs solar cells have been developed for this application during the last decade. The silicon cells selected for the flight programme have efficiencies of more than 26% at deep space conditions.

Other important European satellite projects with major solar array subsystems are the Polar Platform with its 8 kW single-wing array, the Artemis Telecommunication Satellite and the ATV, the Space Station transfer vehicle. Eureca (European Retrievable Carrier) is the first European spacecraft designed to return to ground after operating for just a year in low-Earth orbit (LEO, with altitude usually less than 1000 km). Olympus is a large telecommunications satellite designed for a planned ten-year lifetime in geostationary orbit (GEO, at altitude of about 36 000 km). Agora (Asteroid-Gravity Optical and Radar Analyser) is a scientific mission which aims to explore the asteroid belt at a distance of 2.5 astronomical units from the Sun. Columbus is Europe's space station related concept which is currently in preparatory programme phase.

5.6.2 Satellite PV system

In common with their terrestrial counterparts, the satellite PV system is usually divided into the photovoltaic generator, the storage subsystems and power conditioning which also includes distribution.

Most satellites require energy storage for two reasons:

- During the eclipse period, the solar array provides no power and at least certain vital equipment has to be supplied with electricity from batteries which are recharged in the sunlit part of the orbit. In some cases, the full eclipse power for the payload is required.

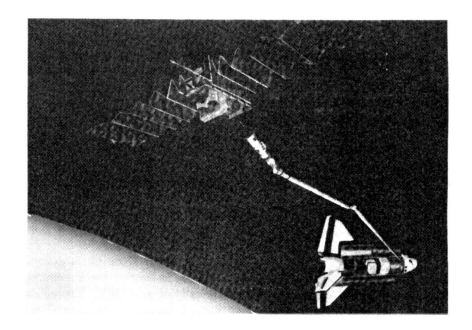

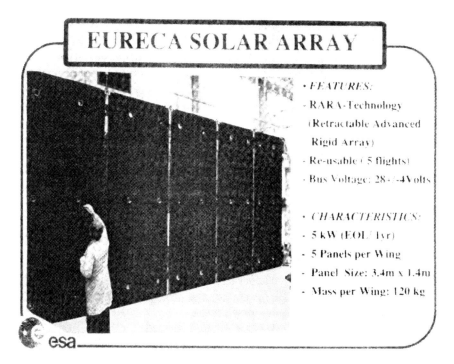

Fig. 5.17 The Eureca PV array (courtesy of the European Space Agency)

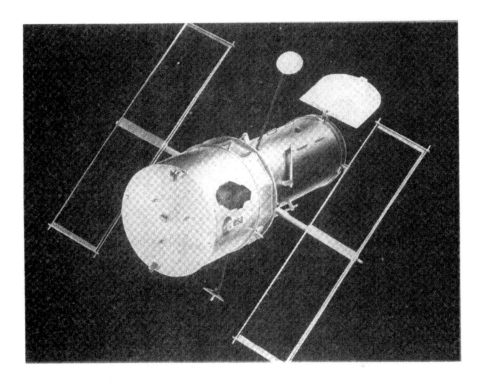

HST SOLAR ARRAY

• *FEATURES:*

- FlexibleBlanket Double Roll-up [Retractable]

- In-orbit replaceable/ EVA/ Jettison

- Bus Voltage:34 Volts

• *CHARACTERISTICS:*

- 4.65 kW (EOL/ 4 yrs)

- 4 x 5 SPA's

- SPASize:2.4m x 1.08 m

- Mass per Wing: 152 kg

esa

Fig. 5.18 The Hubble space telescope array (courtesy of the European Space Agency)

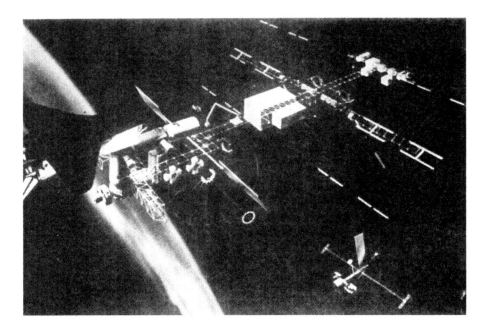

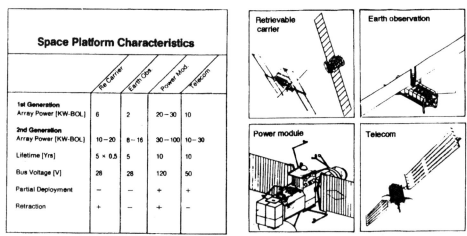

Fig. 5.19 The space station (courtesy of the European Space Agency)

- For peak power requirements which exceed the solar array capacity, batteries are required to act as an energy buffer.

The output power from the satellite's PV generator changes with the orbital position of the satellite, the season, and the age of the system. The power-conditioning subsystem provides the necessary regulation before the power is distributed to the various users, and also controls the charging and discharging of batteries.

5.6.3 The PV generator

The most obvious requirement on the design of a satellite solar array is its absolute power level, resulting from the specific mission requirements. The power-to-mass ratio is another important factor reflecting the importance of a lightweight design dictated by the very high cost of launching a kilogram of spacecraft into orbit. The evolution of solar array technology with respect to these two parameters is shown in Fig. 5.21. Of high importance in the array design are also the environmental conditions which vary considerably from mission to mission, and are discussed in more detail below.

The space solar array must generally have good resistance to the particle and UV irradiation. The particle flux is particularly damaging in orbits which pass through the van Allen belts of electrons and protons trapped by the electromagnetic field of the Earth.

Satellites in a low-Earth orbit move in a relatively benign radiation environment but face other hazards due to residual atmosphere. The ionised plasma causes power loss in high-voltage arrays due to leakage, as well as problems associated with electric discharge. The presence of atomic oxygen causes erosion of exposed surfaces, and large solar arrays experience air drag.

Fig. 5.20 (a) The Rosetta satellite on display in London (courtesy of the European Space Agency)

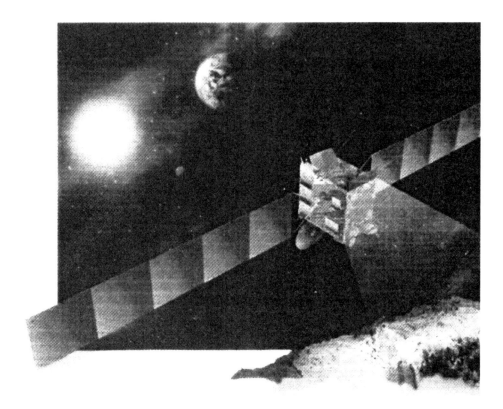

Fig. 5.20 (b) Artist's impression of the Rosetta mission (courtesy of the European Space Agency)

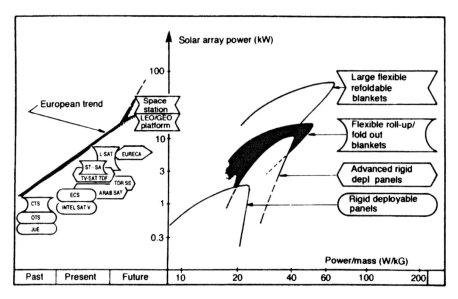

Fig. 5.21 Evolution of space solar arrays

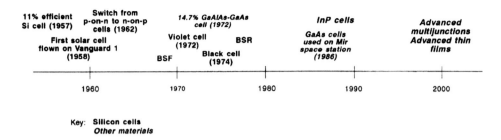

Fig. 5.22 Historical development of space solar cells

Another aspect of the orbital missions, particularly important for arrays designed for long life, are the Sun/eclipse thermal cycles which may cause fatigue of solar cell interconnections.

The space solar cells, therefore, must have high operation efficiency, high resistance to irradiation by energetic particles (mainly electrons and protons) and a good power-to-weight ratio. The historical development of these cells is summarised in Fig. 5.22.

Following the early Si cells based on n-type substrates and p-type diffused emitters, the cells in use since the early 1960s have been based on p-type substrates because of their higher radiation resistance. Subsequent developments include the incorporation of the back-surface field (BSF), back-surface reflector (BSR), the use of a shallow emitter (the 'violet cell'), and top surface texturing (the 'black cell').

Monocrystalline silicon solar cells have been further improved towards higher efficiency and more particle radiation-resistant devices. High efficiency silicon (hi-eta) cells with 16% beginning-of-life efficiency and more are now available at reduced costs. However, the monoculture of silicon cells has, in the last few years, been replaced by a variety of different cell concepts and materials. GaAs-on -Ge monojunction cells have been successfuly used in many applications, boosting the efficiency to above 19%. Multijunction GaInP/ GaAs cells have been in production for more than a year, with efficiencies between 22% and 24%. The on-going development aims at triple- and quadruple-junction cells aiming at efficiencies above 30%. In parallel, the use of advanced thin-film cells (e.g. CIS, CdTe and amorphous silicon) is considered for low-cost commercial applications.

Summary

Photovoltaics has been the principal means of electricity supply to satellites since the birth of the space era. Various satellite power systems have been examined. The criteria placed on space PV arrays have been discussed in detail, alongside the structure of the complete system. A brief historical review of space solar cells has been presented.

5.7 GRID-CONNECTED SYSTEMS

5.7.1 Introduction

Grid-connected systems are now being seriously considered to supplement the conventional power generation in many industrialised countries. Indeed, there are already many grid-connected demonstration projects which study this possibility. Although they have yet to become viable on economic grounds, the participation of PV in large-scale power generation is viewed with increasing prominence as a means of halting the adverse environmental effects of conventional energy sources (see Chapter 6).

Two types of grid-connected installations will be reviewed in this section: centralised large-scale PV power generation in PV power stations, and an alternative, distributed form of power generation in units located directly at the customer's premises.

Box 5.4 What is grid connection?

Public electricity supply has developed in all industrialised countries of the world. Whether in private or public ownership: it has evolved in a similar way, and comprises generation, transmission and distribution.

- Generation is in large centralised power stations.
- A high-voltage transmission system transmits the energy to large substations.
- A local (low-voltage) supply network distributes the energy to the end user.

The fundamental parameters are also remarkably similar. Alternating three-phase current, frequency 50 or 60 Hz, transmission voltage between 400 and 800 kV, distribution voltages between 11 and 132 kV, and a supply voltage of 110–240 V.

Because the purpose of the high-voltage transmission system is to interconnect all points of generation (power stations) and distribution (substations) it has the appearance of a grid, giving rise to the popular name for the transmission system (Fig. B5.8). The lower voltage distribution systems are supplied by the transmission grid and have had little in-built generation. Hence, they are rarely interconnected and are normally radial networks.

The privatisation of the system and liberalisation of the markets has been accompanied by many changes. One of these changes is the growth of so-called *distributed* or *embedded generation*. This description is given to smaller generators connected at lower voltages than the transmission voltage, often to the distribution network. Additional reasons for the introduction of embedded generation include the growing interest in the renewables (including

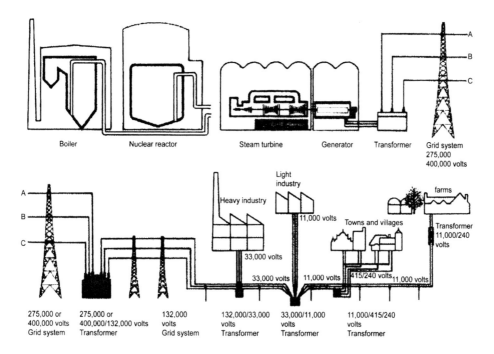

Fig. B5.8 A schematic diagram of the generation, transmission and distribution system (Source: CEGB)

photovoltaics) and the application of co-generation (Combined Heat and Power, or CHP).

The ability to connect any electricity generator to the supply network greatly enhances its usefulness, and photovoltaic generators are no exception. There are a number of issues concerning the connection of photovoltaic generators to the supply network:

• Reinforcement of network in relatively underdeveloped areas,
• Assistance with voltage support on long rural feeders,
• Displacement of less environmentally friendly generating methods.

It is into the low-voltage distribution network that the majority of photovoltaic generators will be connected. Their size (most frequently between 1 and 100 kW) is usually accommodated adequately on the 440 V supply network. However, because this network is normally radial in nature and not interconnected, the connection may at times present some problems. Although the term 'grid-connection' is now firmly established it is worth remembering that photovoltaic generators are rarely connected to the high-voltage 'grid'.

(Contributed by R. J. Arnold)

5.7.2 PV power stations

A PV power plant feeds the generated power instantaneously into the grid by means of one or more inverters and transformers. Most larger systems use line-commutated inverters equipped with a maximum-power-point tracker (Fig. 5.23(a)). Figure 5.23(b) shows the first PV power station, built at Hysperia in southern California in 1982 with nominal power specification 1 MW. It uses crystalline silicon modules mounted on a two-axis tracking system.

(a)

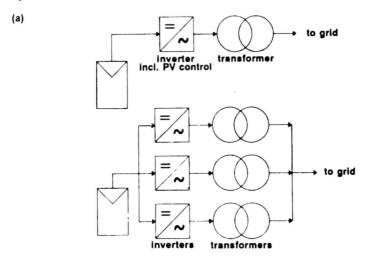

(b)

Fig. 5.23 (a) Configuration of PV power plant with one or several inverters; (b) PV power station (courtesy of P.D. Maycock, Energy Systems Inc.)

There is little doubt that, at present, the cost of generated electricity from PV plants is not generally competitive with the conventional power sources: the cost of electricity from PV power plant is typically 10–15 times higher than from a large thermal plant. Additional consideration must also be given to the area of the system as, for example, a 25 kW plant represents an active PV array surface of approximately 250 m^2, and occupies a land area of roughly 1000m^2.

There are, however, some exceptions. It may already be economic to set up a PV plant to assist the local grid during periods of peak demand rather than construct a new power station. This is known as *peak shaving*. It can also be cheaper to place a small PV plant at the end of a transmission line rather than upgrade the line so that it can take more power from the grid (*embedded generation*, see Box 5.4).

The owner/operator of a grid-connected system will usually need to obtain permission from the local electricity distribution company for operating in parallel with the public electricity supply. The utility will wish to be satisfied that the generator does not prejudice the safety of personnel or protection of equipment connected to the network. The inverter needs to have adequate loss-of-mains protection which ensures that it will discontinue operation when the public supply is shut down. Continued operation in the absence of utility power (islanding) is frequently a cause for concern when a number of embedded generators operate on the same section of the distribution network. The power produced by the inverter must also be of sufficient quality, and the utility is likely to place restrictions regarding flicker, power factor and harmonic distortion.

PV power plants are discussed further in Chapter 7 where we consider large-scale PV projects.

5.7.3 PV in buildings

PV arrays mounted on roof tops and facades offer the possibility of large-scale power generation in decentralised medium-sized grid-connected units. The PV system supplies the electricity needs of the building, feeds surplus electricity into the grid to earn revenue, and draws electricity from the grid at low insolation (Fig. 5.24). This potential type of PV operation within the power generation sector is presently favoured in many industrialised countries. Indeed, the integration of photovoltaic modules in buildings is now seen as part of a new architecture, often in conjunction with other energy efficiency and conservation measures, where aesthetic considerations often outweigh the economics of power generation (see Box 5.5).

It has been estimated in several studies (Germany, UK, Holland, and in Switzerland) that the technical potential of PV arrays installed on roofs and facades of buildings approaches the capacity needed to supply the total electricity consumption. For example, Bloss *et al.* (1991) estimate that in Germany there are about 650 km^2 of roof area which is technically suitable for PV installations. Assuming 8% conversion efficiency this represents a distributed

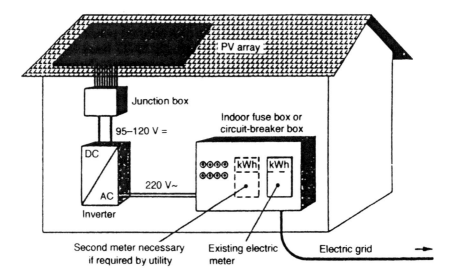

Fig. 5.24 Roof-top grid-connected PV system

power plant of $52\,GW_p$. This value should be higher still if the potential of facades were included. Caution should be exercised when using these estimates, however, as they do not take into account the variable nature of both the electricity demand and solar energy supply.

A large number of roof-mounted PV systems have been installed in Germany in the early 1990s as part of the government sponsored 'Thousand Roofs Programme' (see Box 5.6). Based on the experience gained from this pilot scheme, much larger programmes have recently been announced or are already in operation (for example, the Japanese 70 000 Roofs Programme, US 1 Million Roofs, German 100 000 Roofs or European Union 500 000 Roofs). Fig. 5.25 shows the example of a residential roof-mounted PV array in Amersfoort, near Amsterdam.

The main advantages of these distributed systems over large PV plants are as follows:

- There are no costs in buying the land and preparing the site,

- The transmission losses are much lower because the load is on the same site as the supply.

- The value of the PV electricity is also higher because it is equal to the *selling price* of the grid electricity which has been replaced, rather that to the cost of generating it.

However, it should also be noted that the price paid by utility companies for electricity exported from a decentralised source is not the same as the utility sale price of electricity. In some countries, utilities do buy PV electricity at a similar price to which it is sold. In other countries, however, PV electricity is bought at

Fig. 5.25 Photovoltaic solar roofs

a considerably lower rate. The optimum economic benefit is then derived by consuming all PV-produced electricity, with direct reduction of the energy imported from the utility. Thus grid-connected PV systems are ideal for loads which vary in proportion to the irradiation. Typical loads are air-conditioning, refrigeration and pumping. Other significant loads can be timed to operate when PV power is likely to be available. Examples include washing machines and clothes dryers which can operate on timing clocks.

Box 5.5 Building Integrated Photovoltaic Systems

Photovoltaic modules are being architecturally integrated into low-energy buildings designed with multifunctional outer skins that offer protection, electricity generation and aesthetics. These buildings have the potential to become an important part of the built environment, and can serve as independent power stations that generate electricity for local use as well as social, economic and environmental benefits. Building Integrated Photovoltaic Systems (BIPVS) belong to the concept of 'building as an energy system', where architectural elements have the multiple function of complying with various building requirements which include control of energy flows.

The majority of electricity used in commercial buildings is used during the day, when there is good correlation between electricity supplied by the solar system and the building's demand for electricity. The system is usually connected to the electricity grid and surplus electricity produced by the PV system can therefore be exported to the grid and need not be stored in batteries as in stand-alone systems.

Multifunctionality

Multifunctionality of the PV module in buildings is a key property for its implementation in buildings. BIPVS should combine with other low-energy strategies such as shading, natural ventilation and daylighting, to upgrade its value as a product that competes with conventional building elements in the market. By reducing the energy demand required to light and to control temperatures in the building, the electricity produced by the PV modules becomes a higher proportion of the electricity provided to the building, reducing its technical and environmental impact.

The multifunctional feature of PV modules integrates well with other sustainable technologies in the building's energy strategy. The photovoltaic system may then form part of a complete energy system that combines active and passive elements as a whole. Some of these sustainable technologies include active solar thermal water heating, energy efficient appliances, good thermal insulation, efficient heating, ventilation and air conditioning (HVAC) equipment, and advanced glazing.

Modules for Building Integration

As the market for photovoltaics in the built environment is growing (estimated at close to 25% yearly, see Box B5.1), photovoltaic products have been developed to create more attractive aesthetics in modules without significantly reducing their efficiency. The different types of photovoltaic modules — for example, monocrystalline, multicrystalline or amorphous silicon— have different aesthetic and functional properties. Roof tiles, laminates, glass–glass modules, and other configurations that adapt in different ways to the outer skin of buildings have been created in different shapes and sizes to match architectural design specifications. Solar cell components, including contact grids and antireflection coatings, have been modified to change the visual appearance, colour and texture of the cells to achieve better aesthetics. Photovoltaic modules for building integration should be easy to install and maintain so that a high-level of technical expertise is not needed during the installation or operation of the system.

Subsidies and Costs

Aesthetics and utility are usually more important than price when selecting materials for building facades. In the case of BIPVS, the main obstacle is neither technology nor price but an adequate financing mechanism is an important part of the overall project. Most buildings with architecturally integrated photovoltaic systems have been partly financed by public subsidies, or belong to companies that are associated with photovoltaic industry such as local electric utilities or construction companies. Germany, Switzerland and the Netherlands are among European countries that offer the most promising financing schemes. Many analysts believe that photovoltaic prices will halve in the next ten to 15 years. It is important to note that BIPVS are not attractive because of their payback time frames as is sometimes mistakenly believed by architects and planners. The true value of these systems is in design and environmental consciousness. According to a study made by the Institut für Energieversorgungs-technik, BIPVS prices now range between £620 and £790 per square metre, which could compete with other expensive facade elements, such as polished stone. Generally, the photovoltaic system will add between 2 and 5% extra cost to total construction costs of commercial buildings.

These concepts are illustrated below using three examples of actual building-integrated PV systems.

Example No. 1: The Academy of Mont Cenis Herne
(Figs B5.9—B5.11)

This 12 000 m² building has been erected on the site of the former Mont Cenis coal mine in Herne-Sodingen in Germany's Nordrhein-Westfalen and is part of the Internationale Bauausstellung Emscher Park. It is a controlled Mediterranean microclimate enclosed by a glazed envelope that measures 180 × 72 × 16 metres.

Fig. B5.9 The Mont-Cenis Academy

Fig. B5.10 Mont-Cenis Academy: a view of the interior building, with a shading pattern created by the solar cells

Fig. B5.11 Mont-Cenis Academy: the roof PV modules. Note the different density of solar cells creating a 'cloudy sky' pattern

The photovoltaic modules integrated in the roof of the glass structure form cloud patterns, and provide shading and protect from glare and direct solar radiation. The density of solar cells in the modules varies from 58% to 86%. The energy rating of the panels therefore varies from 192 to 416 W_p. Photo voltaic modules are also incorporated into the west façade of the envelope. Six hundred inverters transform the DC current into AC which can be exported to the local distribution system. The annual energy generated (about 750 MWh) is about twice the requirement of the building, and will also be used to supply electricity to 200 new households.

The energy system at the Mont-Cenis Energy Park combines the photovoltaic generator with other environmentally friendly energy sources. The former mine pitheads on the site release more than one million m^3 of gas per year which contains 60% methane. The gas fuels two co-generation units which produce both electricity and heat. The electricity is fed into the public supply system and the heat is used for the academy, surrounding housing and a nearby hospital. The usage of the mine gas avoids the release of methane into the atmosphere and, by providing energy, reduces CO_2 emission by 12 000 tonnes a year.

About 80% of the solar electricity will be produced during the summer period. In winter, most of the energy needs will be supplied by the co-generation units. The combination thus helps to even out the seasonal imbalance which is often encountered in PV systems in northern locations. The energy system is further supplemented by a 1.2 MWh battery which stores the electrical power, balances fluctuations in the AC energy and reduces peak load.

Start of Operation: March 1999
Project partners: State of Nordrhein-Westfalen, Pilkington Solar, architects Jourda et Perraudin and HHS Planer + Architekten
PV System Cost: DM 14.7 million, DM 15 per watt. 49% financed by Nordrhein-Westfalen and EU, 51% financed by local utility company.

(*Adapted from the project brochure available at the Academy and from Renewable Energy World, May, 1999.*)

Example No. 2: Doxford International Business Park, Sunderland, UK (Figs B5.12—B5.14)

A 4600 m² three storey building at the Doxford International Business Park adopts a holistic energy strategy. The main objective of the environmental design was to find a symbiosis between the low-energy measures and those needed for an effective photovoltaic installation.

Two arrays of total 352 PV modules of area 532m² are integrated into the south-facing façade which incorporates alternate bands of modules and conventional glazing. Each array has 17 strings forming four sub-arrays. Up to five strings are connected in parallel in array boxes feeding a 35 kW central inverter. Two single strings in the centre of the façade are each separately connected to 850 W string inverters.

Fig. B5.12 Doxford building: front elevation (courtesy of Akeler Developments)

Fig. B5.13 Doxford building: view from inside (courtesy of Akeler Developments)

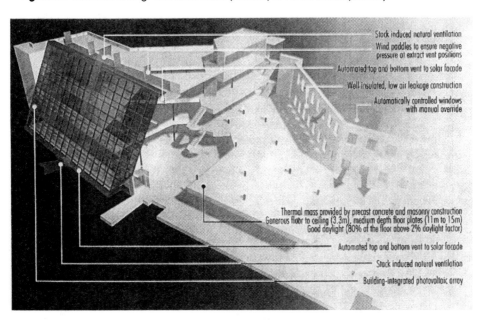

Stack induced natural ventilation
Wind paddles to ensure negative pressure at extract vent positions
Automated top and bottom vent to solar facade
Well insulated, low air leakage construction
Automatically controlled windows with manual override

Thermal mass provided by precast concrete and masonry construction
Generous floor to ceiling (3.3m), medium depth floor plates (11m to 15m)
Good daylight (80% of the floor above 2% daylight factor)
Automated top and bottom vent to solar facade
Stack induced natural ventilation
Building-integrated photovoltaic array

Fig. B5.14 Doxford building: scheme of air flow (courtesy of Akeler Developments)

The façade works not only as an electricity generator but also as an insulation system and it contributes to office ventilation and cooling. Vents at the bottom and top of the façade support the natural stack effect, with hot air rising and passing out of the building. The vertical air plume assists in pulling fresh air across office floors from open windows on the building's north side. This cooling effect can also lead to increased electricity generation as reduced module temperature increases the module efficiency (see section 4.3).

Testing by the Institut für Energieversorgungstechnik in Kassel in Germany has shown that the additional gain from ventilated cells was higher than the energy used to run the vents. It is anticipated that the building will have an energy demand of $85\,kWh/m^2$-year compared with $250\,kWh/m\}^2$-year for a well-designed air-conditioned office building.

Date of Completion: April 1998
Project partners: Akeler Developments, Studio E Architects, Schueco
PV System: $73\,kW_p$ powering 30% of building annual energy needs.
Project cost: European Regional Development Funding grant £1.5 million for the incorporation of PV and other low-energy features.
Expected Annual Energy Generation: 55.1 MWh, reducing annual emission of CO_2 by 375,600 kg

(*Adapted from New Review, Issue 37, August 1998. Published by ETSU, Harwell, UK. Further information available at* **www.akeler.com/doxford/ schuehle.**)

Example No. 3: G8 Solar Showcase, Centenary Square, Birmingham, UK (*Fig. B5.15*)

A $113\,m^2$ pavilion was designed and built in six weeks to show the potential of photovoltaic cells in domestic and commercial buildings at the annual 1998 G8 summit in Birmingham. It demonstrates how energy-saving practices can be used to create low-energy housing or offices. The building is a prototype in the sense that its orientation and the level of glazing and insulation can adapt to suit a particular function and location.

Two heat recovery systems are employed. One uses the heat from the inverters and from the rear of the PV panels. A fan blows this hot air through ducts into the damper flap near the entrance from where it is directed into the interior or exhausted outside, according to the internal heating requirements.

The second system draws hot air from the top of the pavilion through a secondary system of ducts again into the damper next to the entrance. If the heat is not needed internally, the hot air is directed to an underfloor chamber which contains a 'pebble bed', providing an additional thermal mass to the structure and helping to minimise diurnal and seasonal temperature fluctuations.

Start of Operation: 14 May 1998
Project partners: Arup Associates, BP Solar and Pilkington.
PV system: $15\,kW_p$ powering ventilation, lighting and electrical equipment.
PV array: 176 BP Saturn modules arranged in a curved and angle shape, and connected in 22 parallel strings, each feeding an inverter.
Total Project cost: £250,000
Expected Annual Energy Generation: 12.0 MWh

(*Adapted from Renewable Energy World, July 1998.*)

Fig. B5.15 G8 Solar Showcase (courtesy of BP Solarex)

References and Bibliography

Renewable Energy World. Various Issues in 1998 and 1999.
New Review. Published by ETSU, Harwell, UK.
Building with Photovoltaics. Published by Novem, The Netherlands Agency of Energy and the Environment, P.O. Box 8242, 3503 RE Utrecht, The Netherlands.
Integration of Solar Components in Buildings. Published for the European Commission by Institut Català d'Energia; Av. Diagonal, 453 bis, àtic; 08036 Barcelona, Spain.
www.akeler.com/doxford/schuhle
Entwicklungsgesellschaft Mont-Cenis (Brochure with description of the project) Stadt Herne/Montan Grunstuckgesselschaft (1998)
(*The manuscript 'Building Integrated photovoltaic Systems' was prepared by Rogelio Leal*)

A diagram of a typical domestic PV system is shown in Fig. 5.26. It consists of a photovoltaic generator connected via a DC protection and isolation unit to the inverter. The inverter supplies power to the domestic distribution board, or the surplus to the utility grid, via two meters which measure the import and export of power from the utility network. The power generated by the PV system is usually maximised by employing maximum-power-point tracker which is normally included as part of the inverter.

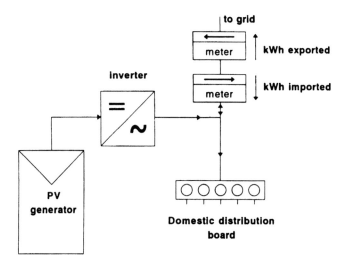

Fig. 5.26 Configuration of a residential grid-connected system

Box 5.6 The German 1000 Roofs Programme

The '1000 Roofs Programme' was initiated jointly by German Federal and State Governments on 23 September 1990. After the reunification of Germany, the programme was extended to include the new state as well. Interest in the programme was overwhelming. The German Federal Ministry for Research and Technology (BMFT) had to send out information about photovoltaic systems to a total of 60,000 citizens. As Dr Walter Sandtner, head of the BMFT department responsible for the programme remarked: 'On that basis, the German population is probably the best informed in the whole world'.

From the 4000 applications for the programme that were submitted, 2250 systems were approved according to the distribution which was agreed with the individual states: 100 systems in each of the City States, and 150 systems in each of the other States. BMFT subsidised 50% of the costs in the western States, and 60% in the east. The States made an additional contribution of 20% and 10%, respectively, bringing the total subsidy to 70%, with the exception of Saarland which considered 50% subsidy to be sufficient.

The 1000 Roofs Programme was conceived as a demonstration programme meaning that only those systems would be funded which could act as a prototype for others. Furthermore, it was clear that the DM80 million which was set aside for the project would not be sufficient to fund all the systems for which applications were submitted. For this reason, the programme had to narrow down to systems with peak power between 1 and 5 kW$_p$ which were to be mounted on the roofs of single or duplex houses. Systems with battery storage were not included in the programme. System operators also had to agree to participate in a five year monitoring programme. A wide variety of locations, module manufacturers, trades companies and electricity utilities were targeted to ensure the widest impact of this demonstration programme.

The programme aimed to achieve four principal goals:

- to promote architectural and construction integration of the PV systems in the roofs (Fig. B5.16);
- to stimulate electricity saving by the users who were encouraged to adapt their consumption to the pattern of solar generation (Fig. B5.17);
- technical optimisation of all components;
- to gain know how in the installation of roof mounted grid connected PV systems.

These goals were largely achieved. Although at the beginning the programme provided a demonstration of a wide variety of architectural approaches – some more appealing than others – towards the end of the programme the system designers learned to take into account the importance of aesthetic aspects. A subsequent sociological study is under way to assess the impact of PV systems on the electricity consumption by the owners/operators. After some teething troubles, it is now routinely possible to install a PV system on a roof anywhere in Germany within a few days.

The success of the programme has been confirmed by similar programmes which were started in other countries, notably Austria, the Netherlands and Japan. It would be a pity if the gratifying upswing in solar generated electricity which the 1000 Roof Programme has stimulated, concluded Dr Sandtner, were not continuously extended and supported.

Fig. B5.16 Winner of the competition for the most attractive PV system within the 1000 Roofs Programme (photo courtesy of Frauenhofer Institute for Solar Energy Systems)

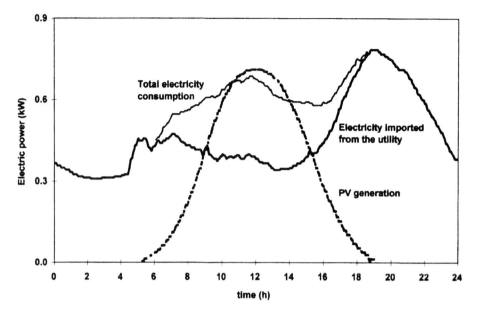

Fig. B5.17 Average daily power profile (courtesy of Frauenhofer Institute for Solar Energy Systems)

Year	1994	1995	1996
Number of systems	1203	1332	1182
Consumption (kWh/year)	4426	4525	4783
Solar fraction (%)	52	53	49
Direct consumption (%)	46	46	47

(Adapted from: *Solar Electricity from a Thousand Roofs*, published by Frauenhofer Institute for Solar Energy Systems)

Summary

Photovoltaics is now being seriously considered as a potential large-scale source of electric power in grid-connected applications. Two principal types of such systems have been discussed: PV power stations and a decentralised range of PV systems in buildings. The latter form has been highlighted by giving a decription of the potential scale of applications, the system structure, and its interaction with the utility.

SUMMARY OF THE CHAPTER

A range of photovoltaic applications has been examined. Most of these were stand-alone small- to medium-sized systems which exploit the reliability of PV power supply and require little maintenance, making them suitable for remote or isolated locations.

Economic analysis based on life-cycle costing has been analysed in detail and applied to PV systems. A comparison with alternative electricity sources in rural locations has been made, with the conclusion that PV systems provide a good option in a range of applications which require grid extension or small diesel generating systems.

Rural PV electrification has been discussed with emphasis on the social benefits that this brings to rural populations. Particular attention has been given to solar pumping where PV systems have been applied with success for some time.

Professional PV applications have been described, and their special features highlighted. Space PV systems, an important application of PV systems since the early days of modern photovoltaics, have been discussed in some detail.

In conclusion of the chapter, an overview has been given of potential grid-connected systems that are under consideration for large-scale power generation. Both centralised and decentralised types of such systems have been discussed.

BIBLIOGRAPHY AND REFERENCES

BOGUS, K. Space photovoltaics–present and future, *ESA Bulletin* 41, 70–77.

HILL, R., *Applications of Photovoltaics*, Adam Hilger, Bristol, 1988.

MARKVART, T. Radiation damage in solar cells, *Journal of Materials Science: Materials in Electronics* **1** (1), 1990: 1–8.

MAYCOCK, P. D. and STIREWALT, E. N. *A Guide to the Photovoltaic Revolution*, Emaus, PA., 1985.

BLOSS, W. H., PFISTERER, F., KLEINKAUF, W., LANDAU, M., WEBER, H. and HULLMAW, H. Grid-connected solar houses, In: *Proc. 10th European Photovoltaic Solar Energy Conf., Lisbon*, 1991: 1295–1300.

SELF-ASSESSMENT QUESTIONS

PART A. True or false?

1. PV can already be an economically viable source of electricity for rural electrification.

2. The annualised life-cycle cost of a PV system is the total life-cycle cost divided by the number of years when the system will remain in operation.

3. PV-powered remote monitoring systems can only be installed in locations with high solar radiation.

4. Domestic PV supply is usually designed for lower reliability than PV-powered telecommunication systems.

5. Multicrystalline silicon cells are a frequent choice for space applications because of their low cost.

6. A major advantage of PV systems is their reliability.

7. A large component of the cost of a PV pumping system is the battery cost.

8. PV can potentially provide a considerable part of the household electricity supply in many industrialised countries.

9. PV provides a good source of electricity for water heating.

10. The largest PV power station in the world has 150 MW of installed power.

PART B

1. Name some applications where the installation of a PV system is fully justifiable on the grounds of conventional economic arguments.

2. What advantages do PV systems have in comparison to diesel pumping?

3. Outline the demand characteristics of: (a) village water supply; (b) irrigation pumping.

4. What are the principal requirements on space solar cells? What are the principal hazards that affect the satellite array design?

5. What are the main advantages of distributed PV systems over centralised power stations? What is the general configuration of the system?

PART C

1. You are setting up a PV-powered battery supply. You have the choice between a battery costing $500 with a life of 5 years, or a battery costing $1100 guaranteed to last 15 years. What are the life-cycle costs for each option over 15 years if the discount rate is 10%? Would your choice change if the discount rate was only 5%?

2. A cheap diesel generator costs $500 and has a fuel bill of $100/year. A more efficient generator would cost $800 and cut the fuel bill to $75/year. Both will last 10 years. Which generator would work out more cost-effective at a discount rate of 10% and an inflation rate for diesel of 5%? What is the present worth of all the fuel consumed by each generator?

3. Use the layout of Table 5.4 and a discount rate of 10% to calculate the annualised life-cycle cost of the following system: a 1000 W_p PV array costing $8/$W_p$ and likely to last for 30 years, two batteries costing $400 each that need replacing every 5 years, a power control unit with a 10-year lifetime costing $800, and $50 of wiring that won't need replacing. Assume that you will install the system yourself but that the annual operation and maintenance costs will be 1% of the initial capital cost.

4. If the battery and control unit in the preceding question are 80% efficient, and the array consists of modules of 14% efficiency, what will be the unit electricity cost of the system if it is situated in a region with 5.5 kWh/m^2 day average daily irradiation?

5. A village has a population of 350 people who require clean water, and intend to install a PV pump. Assume a demand of 40 litres per person per day, and that the static water table is 15 m below the surface.

(a) What volume of water per day (in m³) is required?

(b) For two days of storage, what volume (in m³) of tank would be needed?

(c) What is the total head over which water must be pumped? Assume a drawdown of 5 m once pumping begins and that the top of the tank will be 3 m above ground level.

(d) What is the daily hydraulic energy required in kWh/day?

(e) What is the size of PV array needed, in W_p? Assume a daily solar irradiation of 5.0 kWh/m² day, an array mismatch of 0.8, and a daily subsystem efficiency of 0.3.

Answers

Part A

1, True; 2, False; 3, False; 4, True; 5, False; 6, True; 7, False; 8, True; 9, False; 10, False.

Part B

1. For example, remote telecommunication systems, remote monitoring and control systems (seismic recording, climate monitoring, traffic data collection, etc.), signalling systems or cathodic protection.

2. There are no fuel costs or fuel supply problems; PV systems can usually operate unattended, are very reliable, and require little maintenance.

3. (a) Village water demand is fairly constant throughout the year. (b) For irrigation supply, the demand will be very high in some months and almost nothing in others.

4. High efficiency combined with high resistance to particle radiation and high power-to-weight ratio. The principal hazards that affect satellite solar array design are particle and UV radiation, residual atmosphere and ionised plasma in low-Earth orbit, and thermal cycles.

5. There is no cost of buying the land. Value of electricity is higher because it is generated where it is required, and transmission losses are much lower. The configuration is shown in Fig. 5.25.

Part C

1. $1005 and $1100. Yes.

2. The cheaper generator. $781 and $586.

3. LCC = $12 160, ALCC = $1289/year.

4. $0.80/kWh

5. (a) 350 people × 40 l/person/day = 14,000 l/day = 14 m³/day.
   (b) 2 days × 14 m³/day = 28 m³.
   (c) Total head = static level + drawdown + tank height.
       = 17 m + 5 m + 3 m = 25 m.
   (d) Hydraulic energy = 0.0027 × volume × head
       = 0.0027 × 14 m³/day × 25 m = 0.95 k Wh/day
   (e) Array size (kW_p) = hydraulic energy / (irradiation ×F × E) where F = array mismatch factor and E = subsystem efficiency.
       So array size (kW_p) = 0.95 kWh/day/(5.0 kWh/m²/day ×0.8 × 0.3).
       = 0.79 kW_p = 790W_p.

6

Environmental Impacts of Photovoltaics

AIMS

The aims of this chapter are to examine the environmental impacts associated with the production, use and disposal of PV modules, and the methodologies employed in the environmental assessment of PV relative to other forms of electricity generation technologies.

OBJECTIVES

On completion of this unit, the student should be able to:

1. understand the methodology employed in the environmental assessment,
2. assess the environmental impacts of producing PV modules using cells of silicon, copper indium diselenide and cadmium telluride,
3. analyse the hazards associated with the use of PV modules,
4. recommend means of final disposal or recycling of PV modules.

6.1 INTRODUCTION

The environmental damage by energy production has been known for some time, and the assessment of this damage is becoming an issue of ever increasing importance. It is therefore essential that environmental aspects of photovoltaic power generation are considered in some detail.

Written by R. Hill.

The plan of this chapter is as follows. In section 6.2, the requirements on the methodology for assessment are defined, and the concept of external costs is introduced. Section 6.3 discusses the contribution to the external costs in terms of fuel cycles for energy technologies. The environmental impacts of photovoltaic power generation are analysed in section 6.4.

6.2 THE NEED FOR ENVIRONMENTAL ASSESSMENT OF ENERGY SOURCES

It has long been recognised that energy sources are significant contributors to environmental damages. Costs associated with these damages, however, have been treated in general as external to the energy economy since the environment was regarded not as a marketable commodity but as one freely available to everyone. The costs for repairing or reversing environmental damages caused by energy sources were expected to be paid by society at large, e.g. in form of the taxation or via a reduction in the quality of life. By ignoring these costs of energy generation the energy sector based its resource acquisition decisions on information strongly biased against technologies such as renewable sources which impose smaller environmental and social costs than fossil and nuclear sources.

Only by defining and calculating the external costs of all of the different energy technologies is it possible to identify the real costs (internal plus external) associated with the generation of energy. In order to determine these costs a methodological framework has to be developed within which the external costs can be quantified. The methodology has to be simple, easily understandable and applicable to all energy technologies to allow for a comparison of different energy systems and to make sure that all major environmental impacts are internalised into the energy prices on an even footing. The methodology provides the framework within which are outlined the different steps that have to be undertaken to determine how the external costs should be derived, measured and evaluated. In a first step a definition of external costs is given, followed by the identification of the participating energy technologies and their complete fuel cycles, including extraction, processing, conversion, distribution, utilisation and final disposal. Having defined the *fuel cycle* for existing and future technologies their output of pollutants in the environment can be calculated or at least estimated. Based on these pollution outputs the impacts on the environment can be estimated and the associated costs can be derived by various economic valuation techniques. Existing computer models which have attempted to compile process data on the different steps of fuel cycles can be used to facilitate the calculation of the environmental externalities.

Endowed with this information on the total societal costs of each type of energy source, utility regulators and decision-makers have the choice to opt for the optimum energy technology for society and the environment.

Whilst, in theory, this approach seems to work well, in practice it will probably be some time before the incorporation of the state of the environment in economic terms is a realistic option. The attribution of costs to the environmental impacts of energy production and use is still in its developing stages and there are still several externalities which cannot be expressed in monetary, but only in qualitative terms (Fritsche, 1991). However, this fact should not prevent external costs being internalised in resource planning, acquisition and operating decisions as an internalisation will contribute to an improvement in energy decisions generally by sending the appropriate signals to both planners and consumers on how to satisfy the energy needs at the lowest overall costs to society (Bernow *et al.*, 1991). There are strong indications from the recent Geneva Prepcom meetings for UNCED that environmental costs will become an internationally accepted component of traded commodities, including energy.

Summary

The concept of external costs of electricity generation has been defined, and the requirements of a methodology capable of calculating these costs have been outlined.

6.3 METHODOLOGICAL FRAMEWORK

6.3.1 Introduction

In order to quantify external costs of energy technologies a methodological framework must be developed within which the principal issues encountered in performing such an analysis can be addressed. The most significant and decisive part of any external cost analysis is to define, in as detailed a manner as possible, the full set of burdens, impacts and costs associated with different energy technologies and their fuel cycles. A number of criteria have to be considered to provide the methodology with general validity. The methodology should be:

- simple and easily understandable, to allow laymen to make use of the methodological approach,
- adjustable/flexible to allow for modifications and for incorporation of new items,
- applicable to all of the different energy technologies to allow for comparisons of different energy sources on an even footing and to make sure that utility regulators and decision-makers use technologies that are environmentally benign and thus pose the lowest overall costs to society—consistent in its use of evaluation methods of environmental impacts.

6.3.2 External costs

External costs are generally divided into environmental and non-environmental costs. Most of the external costs are grouped under the heading of environmental costs, because they include some of the most significant components of external costs—the valuation of human health and life, acid deposition, greenhouse gases and global warming, to name but the most important ones. There is still a debate about non-environmental costs, about the importance of these costs, their magnitude and the extent to which they ought or ought not to be included in external cost calculation. However, if non-environmental costs are divided into the following four major types: aspect of security, natural resource management, employment and politico-economic instruments, then it is obvious that there are quite considerable societal costs involved and this indicates that non-environmental externalities should not be neglected when calculating the full costs of energy technologies (OECD, 1989).

In order to quantify environmental externalities and link them to the internal costs to derive the total societal costs of these energy systems it is necessary to find a single nominator for all aspects. Monetisation of environmental externalities is one of several valuation approaches that is aimed exactly at this: to express environmental costs in the same monetary unit as used for calculating internal costs of energy technologies (Fritsche, 1991).

Environmental impacts and the associated external costs can be grouped into four categories: negligible costs, small costs, significant costs and large costs. At each stage of the fuel cycle analysis an estimate of the physical magnitude of the environmental impacts can be given by using these four categories. Most notable among external costs are risks of damages to the environment and human health resulting from air pollutants emitted by fossil-fuel-fired generation and from the radiation risks of nuclear plant operations. Environmental costs of air pollutants in general are associated with large cost. The 'greenhouse effect', caused principally by CO_2 emissions from the burning of fossil fuels, and the resulting global warming are only just being assessed but are associated with large costs, even if there are still considerable uncertainties about the order of magnitude of the effects and the costs involved (EC, 1990).

A number of other environmental impacts, such as acid rain, risk to human health and life, and catastrophic events, will incur large costs as well. Impacts such as visual intrusion or operation accidents at the extraction stage or radiation risks may result in quite significant costs, whilst other environmental impacts such as noise or health and safety costs associated with materials inputs and construction can be expected to be small or even negligible (OECD, 1988), provided that care is taken in the design and operation of the industrial processes.

Detailed and reliable information and final figures for the environmental impacts will only become available when the valuation of these impacts has been undertaken.

6.3.3 Fuel cycles, environmental burdens and impacts

A detailed analysis of the fuel cycles of different energy systems is needed to determine the external costs incurred during the various stages of the fuel cycle process and hence to internalise these externalities into the full cost analysis of each of the energy technologies. This in turn facilitates a one-to-one comparison of different competing technologies on an even footing. Based on these data, informed policy actions/choices, either encouraging or limiting particular energy options, can be taken.

The analysis of the fuel cycles of energy systems covers all stages of the energy production process including extraction, preparation, transport, conversion, operation, distribution, utilisation, waste processing and disposal. Each of these stages can impose environmental burdens such as, for example, acid deposition, and these burdens result in environmental impacts, such as forest damage, which in turn pose considerable costs to society (OECD, 1985). The linkage between a particular energy technology and the associated external costs are shown in Fig. 6.1.

Some of the environmental burdens and impacts are well documented and can be identified and measured. In many other cases, however, such as global warming or loss of species, the burden/impact relationships are fraught with considerable uncertainties and non-linearities. These may result from scientific uncertainties, unavailability of accurate data, spatial and temporal variabilities, or the fact that many impacts are the synergistic action between two or more burdens, e.g. the formation of ozone through the photochemical reactions between NO_x and hydrocarbons. Further work will be required to find appropriate methods of defining the burden/impact relationships which cannot be accurately measured so far.

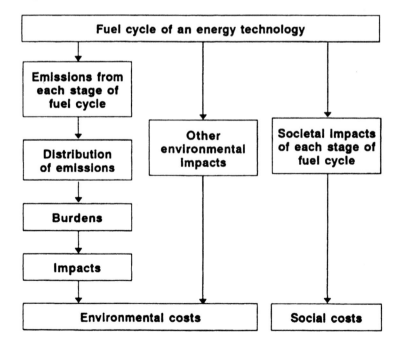

Fig. 6.1 Linkage between fuel cycle and external costs

Summary

The methodological framework for calculating external costs in monetary terms has been considered in detail. The calculation of these costs has been linked to the fuel cycles of the power generation technologies.

6.4 ENVIRONMENTAL COSTS OF PHOTOVOLTAICS

The commercial standard PV module uses solar cells made from wafers of silicon, usually 0.3 mm thick and 10 cm × 10 cm in area. A simplified 'fuel cycle' of silicon solar cells from 'cradle to grave' is shown in Fig. 6.2. The ingots may be either Czochralski-grown single crystals or directionally-solidified multicrystalline cubes. For the latter the third and fourth steps of Fig. 6.2 may be replaced by less costly steps yielding 'solar-grade' silicon rather than the semiconductor-grade material.

Each of the steps shown in Fig. 6.2 has inputs of energy and materials and requires capital equipment, and each also has potential hazards associated with it. The first step is a standard mining operation with associated hazards to the miners and inputs of diesel fuel and machinery. Metallurgical-grade silicon is made in large quantities for the steel industry, with a small fraction going as input to the semiconductor industry. The major emission of this manufacture is of silica dust which can cause lung disease, and there is a substantial energy input. The purification of silicon can involve hazardous materials such as silane, whilst the doping of the silicon involves toxic chemicals such as diborane and phosphine, although only in small quantities diluted in inert gas. These materials are used in the microelectronics industry and their monitoring and control is well established. The materials for construction of the complete PV system, other than the PV modules, are steel, aluminium, copper, concrete and electronic equipment, with which are associated the standard industrial hazards.

The energy used in manufacturing the PV modules and the other components of the PV system is derived from the fuel mix of society and is therefore associated with emissions of greenhouse gases and acidic gases. The energy content of PV modules using silicon wafers has been measured at the Photowatt plant by Palz and Zibetta (1991) as 235 kWh(e)/m^2 for 1990 technology at 1.5 MW$_p$ per annum production rate. Thus for the 1990 crystalline silicon PV technology in small-scale production, the CO_2 emission is around 400 000 tonnes per gigawatt-year of energy output. This compares with the CO_2 output from the most modern and efficient coal-fired plant of 9 million tonnes per gigawatt-year (1991). Boiling Water Reactors have been estimated to emit about 75 000 tonnes of CO_2 per gigawatt-year (Palz and Zibetta, 1991), mainly from the fuel production steps, but this estimate did not include the energy used for decommissioning and waste treatment and storage. The CO_2 emissions estimated for the various current PV cell technologies are shown in Table 6.1, along with an estimate of values which might be typical of 2020 PV technology.

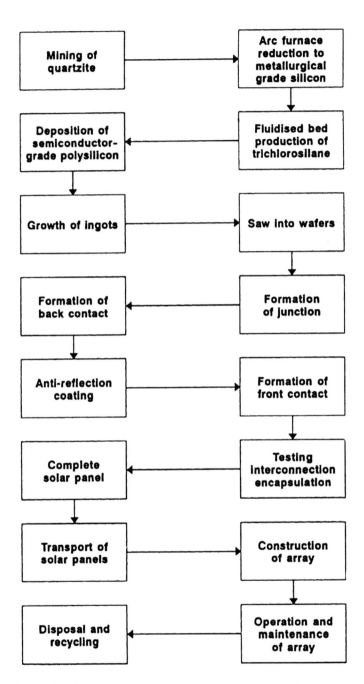

Fig. 6.2 Silicon fuel cycle

Table 6.1 Carbon dioxide emitted in the production of PV modules (in units of kilotonnes per gigawatt-year)*

Cell material	Production scale	Efficiency (%)	Lifetime (years)	CO_2 kt $GW^{-1}yr^{-1}$
Monocrystalline silicon	small	12	20	400
	large	16	30	150
Multicrystalline silicon	small	10	20	400
	large	15	30	100
Thin-film silicon	small	10	20	130
	large	15	30	50
Thin-film polycrystalline materials	small	10	20	100
	large	14	30	40
Future (2020) multijunctions	large	30	30	24

For comparison, the most efficient coal plant emits 9 million tonnes of CO_2 per gigawatt-year, whilst a BWR nuclear plant emits 75 000 tonnes of CO_2 per gigawatt-year (plus an unknown amount for decommissioning and waste treatment).

The thin-film polycrystalline materials have been studied by Hynes *et al.* (1991) and those data in Table 6.1 are taken from their work.

As shown in Table 6.1, there are a number of technologies for PV cells other than those using wafers of silicon. Amorphous silicon cells are widely used in solar calculators, clocks, etc., but seem unlikely to be widely used for power modules. Thin multicrystalline films of silicon on ceramic substrates and thin polycrystalline films of copper indium diselenide (CIS) and of cadmium telluride (CdTe) are in the initial stages of commercial production. An excellent review of the scientific, technical, economic and environmental issues associated with these three materials has been given by Zweibel (1990). Figure 6.3 shows a fuel cycle for CIS modules from Hynes *et al.* (1991).

The environmental impacts of the thin-film silicon cell are similar to those of the wafer silicon cell, but reduced in magnitude because of the smaller volume of silicon used. The CIS cell has a potential hazard in the hydrogen selenide used in its manufacture. A report on CIS manufacture by Moskowitz *et al.* (1990) has concluded that hydrogen selenide can be used safely provided that adequate safety precautions are adopted. There are proposals for the production of CIS which involve the use of solid selenium rather than hydrogen selenide (Badawi *et al.*, 1991) which would largely obviate this hazard.

Both CIS and CdTe cells have a window layer of cadmium sulphide, so both types of cell could potentially present a hazard from cadmium. Cadmium hazards arise in the refining stage, with emissions of cadmium oxide dust, emissions of cadmium during manufacture, or during a fire in which cadmium-containing PV modules were involved, and from possible leaching of

* These 1991 Figures should be compared with the current figures in section 3.

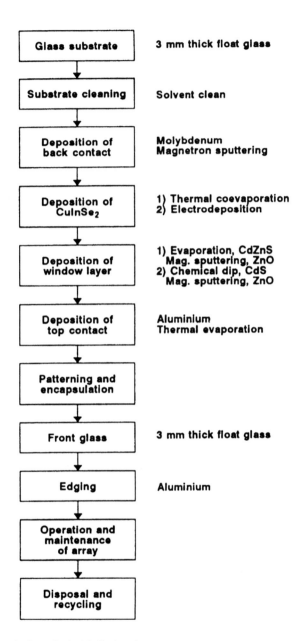

Glass substrate	3 mm thick float glass
Substrate cleaning	Solvent clean
Deposition of back contact	Molybdenum Magnetron sputtering
Deposition of CuInSe₂	1) Thermal coevaporation 2) Electrodeposition

Deposition of window layer	1) Evaporation, CdZnS
	   Mag. sputtering, ZnO
	2) Chemical dip, CdS
	   Mag. sputtering, ZnO

Deposition of top contact	Aluminium Thermal evaporation
Patterning and encapsulation	
Front glass	3 mm thick float glass
Edging	Aluminium
Operation and maintenance of array	
Disposal and recycling	

Fig. 6.3 Copper indium diselenide fuel cycle

cadmium from modules discarded at the end of their working life. In all of these cases, the control strategies are well established in industry, and the magnitude of the hazard is in proportion to the amount of cadmium present in the cells.

In CIS modules, the amount of cadmium used is about $0.04 \, g/m^2$, equivalent to 400 g per MW_p of output. This tiny amount of material is easily controlled during manufacture, whilst cadmium evaporated from modules on a building which caught fire would pose a negligible hazard to anyone far enough from the fire to avoid being burnt to death. Careless disposal of large numbers of CIS modules in one location could pose a hazard from leaching of cadmium and selenium into ground water. However, the rather scarce indium and regulations on the disposal of cadmium and selenium products would lead to recycling of the modules and reuse of the indium, cadmium and selenium.

The PV modules using CdTe cells would contain much more cadmium (about $5 \, g/m^2$) than those using CIS cells, and the potential hazards are correspondingly higher. It is clearly beneficial to use a manufacturing process with a high utilisation of cadmium feedstock, and the success of electrochemical deposition as a production technology gives the opportunity for such a process.

The waste cadmium and tellurium from the manufacturing plant can be economically recycled and emissions reduced to very small levels even for $100 \, MW_p$ per annum production rates. Cadmium dust which might be present in the manufacturing plant could present a chronic hazard to personnel. However, the monitoring and control of such dust is well established in existing industries where cadmium is used, and no additional problems can be foreseen for PV manufacture.

The quantity of cadmium in the CdTe modules increases the potential hazards during operation and disposal compared to those with CIS cells. In operation, the modules can leak cadmium into the environment only if they are broken or subject to high temperatures. Broken modules pose little hazard as the area of cadmium exposed by even a break across the entire module would be less than $3 \, mm^2$, and broken modules would be replaced quickly in routine maintenance. A fire in a building clad with CdTe modules could pose some hazard, although for small areas the hazard would be significant only for those so close to the fire as to suffer far greater hazards from smoke and flames. A large building having a large area of PV cladding could emit cadmium vapour—turning to cadmium oxide dust—into the smoke plume. Moskowitz *et al.* (1990) have studied this issue and conclude that standard fire-safety procedures would protect the population from this cadmium hazard, just as they protect them from the carcinogens in the smoke plume.

Disposal of CdTe modules may need to be controlled more strictly than for CIS modules, and recycling of materials is more important both for environmental reasons and for the value of the cadmium and tellurium. The technological problems associated with this recycling appear to be readily solvable at low cost.

The hazardous emissions from the various different types of solar cells are summarised in Table 6.2.

It is interesting to note that the amount of cadmium contained in the CdTe modules needed to generate energy of, say, 1 GWh over their lifetime is about equal to the cadmium emitted from the smoke-stack of a typical coal-fired station whilst generating the same 1 GWh of electrical energy. Very little of the cadmium in the PV modules would be lost into the environment, so the PV plant is cleaner, even for cadmium emissions, than a coal-fired power station.

Table 6.2 Hazardous emissions from photovoltaics

Material	Production	Operation	Disposal
Silicon	Silica dust Silanes Diborane Phosphene Solvents		
Copper indium diselenide	Hydrogen selenide Cadmium oxide Cadmium dust Selenium Solvents	Cadmium Selenium (in a fire)	Cadmium Selenium (if not recycled)
Cadmium telluride	Cadmium oxide Tellurium Cadmium dust Solvents	Cadmium Tellurium (in a fire)	Cadmium Tellurium (if not recycled)

Note: there is an energy input to all technologies derived from the fuel mix of the nation producing the PV.

Summary

The methodology established in the previous sections has been used to discuss the environmental impacts of photovoltaic power generation. We have discussed the manufacturing technology for cells and modules which have been described in Chapter 3: silicon, and CIS and CdTe technologies.

6.5 CONCLUSION

Photovoltaic systems are almost entirely benign in operation, and potential environmental hazards occur at the production and disposal stages.

Silicon is a very stable material and its release into the environment poses no hazards. In the production of silicon cells the hazards are similar to those encountered in the microelectronics industry, and monitoring and control procedures are well established for even the largest production rates envisaged.

For PV modules based on CIS or CdTe cells, there are potential hazards from the use of hydrogen selenide and cadmium. It is clear from this review, however, that there are well-established methods of monitoring, control and alleviation which can reduce these hazards to within international safety limits for all stages of production, operation to disposal.

All production processes require an input of energy, and this energy is derived from the standard fuel mix of the nation in which production takes place. The production of PV systems has associated with it, therefore, emissions of greenhouse and acidic gases. In the present state of the PV industry, with

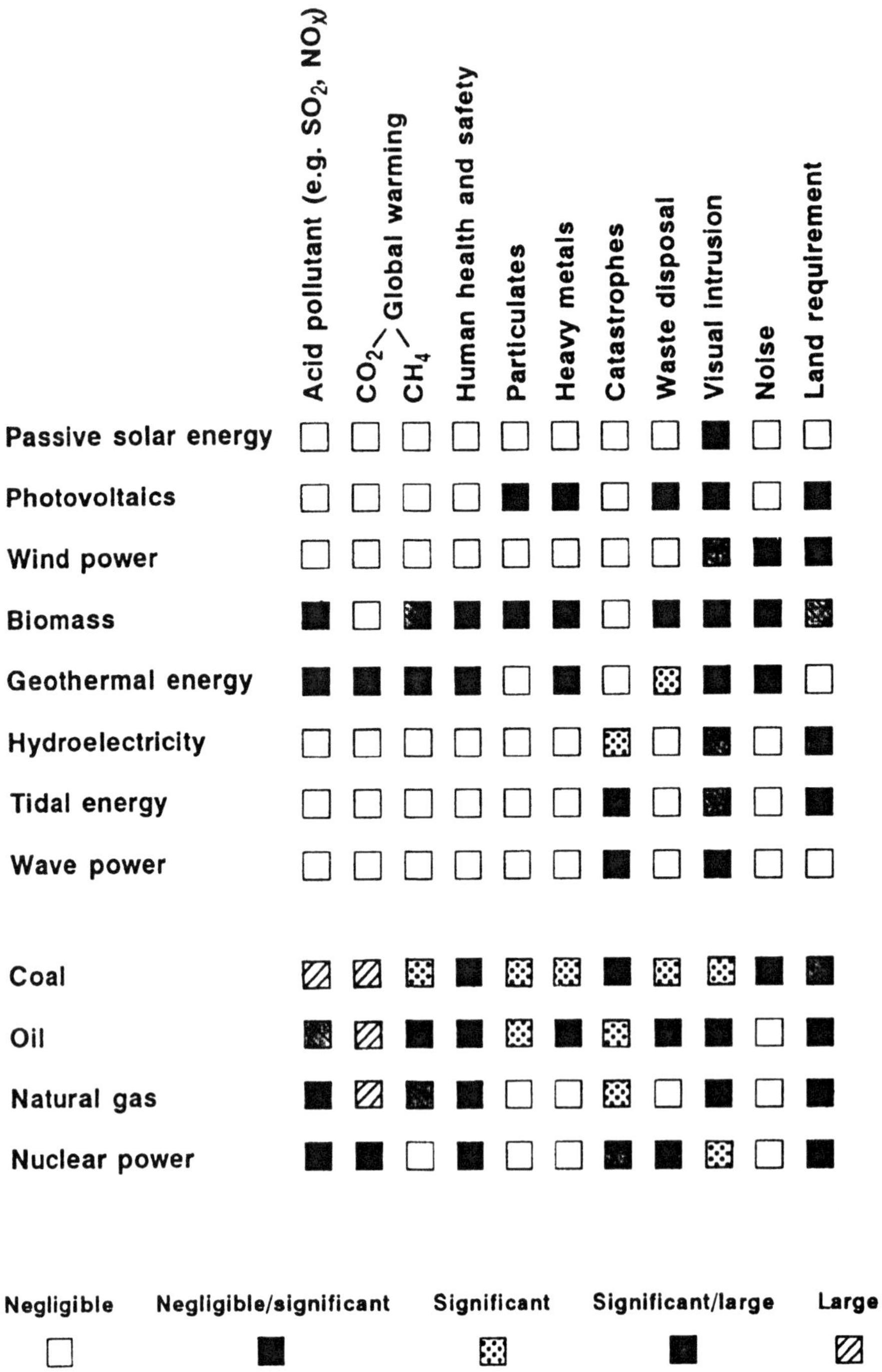

Column headers (left to right):

- Acid pollutant (e.g. SO_2, NO_x)
- CO_2 ⟩ Global warming
- CH_4 ⟩
- Human health and safety
- Particulates
- Heavy metals
- Catastrophes
- Waste disposal
- Visual intrusion
- Noise
- Land requirement

Rows:

- Passive solar energy
- Photovoltaics
- Wind power
- Biomass
- Geothermal energy
- Hydroelectricity
- Tidal energy
- Wave power

- Coal
- Oil
- Natural gas
- Nuclear power

Legend:

Negligible	Negligible/significant	Significant	Significant/large	Large
☐	■	▦	■	▨

Fig. 6.4 Environmental effects of different energy technologies

small-scale production and technologies some way from maturity, these emissions are much less than those from fossil fuels but exceed those from the operation and fuelling of large nuclear reactors. As the scale of production of the new thin-film PV technologies increases, the energy input to PV systems will decrease considerably, with consequent reduction in carbon dioxide emissions, to levels below that of other electricity generating technologies.

Environmental costs arise from many factors, and a valid comparison of the external costs of different energy sources must include all of these factors. There has been a significant effort in Europe and the USA to derive the external costs of electricity generation (Hohmeyer and Ottinger, 1991) and there is a growing consensus on the methodologies used in these calculations and on the external costs associated with the various technologies. Baumann and Hill (1991) have briefly reviewed these methodologies and have produced a matrix which allows a qualitative comparison of the magnitudes of the various components of external costs for the various energy technologies. This matrix is reproduced as Fig. 6.4, and it is clear that PV is the most environmentally benign of all electricity generation technologies presently envisaged for large-scale use throughout the world.

SUMMARY OF THE CHAPTER

The methodology for environmental assessment of energy sources has been reviewed and used to examine the environmental impact of photovoltaic power generation.

The analysis leads to the conclusion that the production and disposal of large PV generators poses significantly lower environmental risks or damages to the environment than the conventional power generation technologies. The hazards during their operation are practically negligible.

BIBLIOGRAPHY AND REFERENCES

BADAWI, M. H., HYLAND, M., CARTER, M. J. and HILL, R., *Proc. 10th EC Photovoltaics Solar Energy Conf.*, Kleuver, Dordrecht, 1991, pp. 883–886.

BAUMANN, A. and HILL, R. *Proc. 10th EC Photovoltaics Solar Energy Conf.*, Kleuwer, Dordrecht, 1991, pp. 834–837.

BERNOW, S., BIEWALD, B. and MARRON, D., *in*: HOHMEYER, O. and OTTINGER, R. L., *eds. External Environmental Costs of Electric Power*, Springer, Berlin, 1991, pp. 81–102.

EC, *Energy and the Environment*, Brussels, 1990.

FRITSCHE, U., *in*: HOHMEYER, O. and OTTINGER, R. L., *eds. External Environmental Costs of Electric Power*, Springer, Berlin, 1991, pp. 191–209.

HOHMEYER, O. and OTTINGER, R. L., *eds. External Environmental Costs of Electric Power*, Berlin, 1991.

HYNES, K. M., PEARSALL, N. M. and HILL, R., *Proc. 10th EC Photovoltaics Solar Energy Conf.*, Kleuwer, Dordrecht, 1991, pp. 461–464.

MOSKOWITZ, P. D., ZWEIBEL, K. and FTHENAKIS, V. M., *Health, Safety and Environmental Issues Relating to Cadmium Usage in PV Energy Systems*, SERI/TR-211-3621, Golden, CO, 1990.

OECD, *Environmental Effects of Electricity Generation*, Paris, 1985.

OECD, *Environmental Impacts of Renewable Energy*, Paris, 1988.

OECD, *Economic Instruments for Environmental Protection*, Paris, 1989.

PALZ, W. and ZIBETTA, H., *Int. J. Solar Energy* **12**, 3/4, 1991.

USDoE, *Energy Systems Emissions and Material Requirements*, Washington, DC, 1989.

ZWEIBEL, K., *Harnessing Solar Power—the Photovoltaic Challenge*, Plenum Press, New York/London, 1990.

SELF-ASSESSMENT QUESTIONS

1. What is the difference between 'direct' or 'internal' costs and 'indirect' or 'external' costs?

2. What are the major components of external costs?

3. What are the linkages between the operation of a technology and its environmental cost?

4. Where do the hazards arise in the production of silicon cells?

5. What issues are raised by the production of thin-film cells?

6. Under what circumstances could PV modules be a hazard during their operational lifetime?

7. Why should modules of thin-film polycrystalline cells be recycled?

8. How does PV compare in its environmental impact with other technologies for generating electricity?

Answers

1. Direct (internal) costs are those which are paid by the company and ultimately by the purchaser. They include energy, materials and equipment used in production, labour costs, and costs of capital. Indirect (external) costs are those incurred by production but not paid by the producer. They are paid by society, or a section of society. Such costs include acid rain, air pollution by transport, the climatic change from the greenhouse effect, etc.

2. Environmental costs affecting the well-being of people and other living creatures and our rural and urban environment, and social costs which affect our well-being by inequitable allocation of resources.

3. *Operation* produces *emissions*. *Emissions* disperse into the environment and cause damage or *Burdens*, e.g. SO_2/m^2. The damage (*Burdens*) causes an *impact* on the environment, e.g. crop yields fall, or forests die. These *impacts* incur costs—the *environmental costs*.

4. During production of silicon from silica—emissions of silica dust. During manufacture of cells—phosphine, diborane, etc. During cleaning and etching operations and disposal of cleaning and etching chemicals.

5. For a-Si, the use of silane, phosphine and diborane. For CIS and CdTe, the use of heavy metals, particularly Cd, and for CIS, the use of H_2Se.

6. PV modules can be an electrical hazard during maintenance in bright sunlight; they can fall on someone whilst being fixed to the array. CIS and CdTe modules could emit Cd during fire.

7. Heavy metals are expensive, scarce and toxic, and should not be put in land-fill sites. They should be recovered and re-used at the end of the module lifetime.

8. Given the small and controllable impact per unit energy output of the PV modules in production, and their almost hazard-free operation and disposal, PV is perhaps the electrical energy generation technology with the lowest environmental impact of all those in use today.

7

Advanced and Specialised Topics

AIM

The aim of this unit is to discuss four individual topics of more specialised nature: large PV systems, photovoltaics under concentrated sunlight, electrochemical photovoltaics and the hydrogen economy.

OBJECTIVES

After completing this unit the student should be able to:

1. understand the specific features of large PV systems, and examine their economic and technical characteristics,
2. explain the advantages and disadvantages of concentrator PV systems, and discuss their operation,
3. to evaluate the photoresponse of contacts between semiconductors and electrolytes as an alternative to conventional solid-state devices,
4. analyse the structure of the hydrogen economy, and show its feasibility and future potential.

NOTATION AND UNITS

Symbol		SI Unit
FF	fill factor	
I	current extracted from solar cell	A
I_{sc}	short-circuit current	A
I_0	dark saturation current	A

Written by A. Sorokin (Large PV projects), P. Davies and J. C. Miñano (Photovoltaics under concentrated sunlight), A. McEvoy (Electrochemical photovoltaics), and M. Specht (Hydrogen economy).

(Contd)

Symbol		SI Unit
F	Faraday constant	C/mol
k	Boltzmann constant	J/K
M	mole, N atoms of molecules of a chemical species	
n	refractive index	
N	Avogadro's number	
P_{loss}	power loss in series resistance	W
q	electron charge	C
R_s	series resistance	Ω
T	temperature of cell	K
V_{oc}	open-circuit voltage	V
V_{DC}	DC bus-bar voltage	V
X	optical concentration ratio	
X_{max}	maximum possible value of X	
G	irradiance	W/m^2
η	efficiency of solar cell	
θ_c	collection angle	
θ_e	angle of tracking error	
θ_{sun}	half-angle subtended by the sun	

Values of physical constants

$$F = 9.65 \times 10^4 \, C/mol$$

$$k = 1.38 \times 10^{-23} \, J/K$$

$$N = 6.02 \times 10^{23}$$

$$q = 1.602 \times 10^{-19} \, C$$

$$\theta_{sun} = 0.27°$$

7.1 LARGE PV PROJECTS

7.1.1 Introduction

Unlike conventional power technologies where large-scale production facilities are generally measured in megawatts, photovoltaics has not yet reached such a large scale of applications.

The world's largest PV plant constructed so far, the Carissa Plains PV power plant in California, USA (Fig. 7.1), reached 5 MW installed peak power capacity. The largest PV plant in Europe, Kobern-Kondorf in Germany, has 500 kW capacity.

At present, the threshold limit value between large- and medium/small-scale PV systems may be set around 25 kW of installed PV peak power. As we have discussed in Chapter 5, this apparently low value in comparison with the traditional power technologies is due not only to the higher cost but also to area considerations.

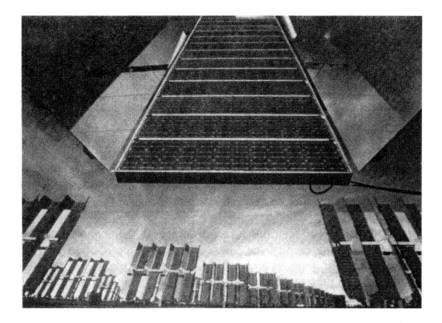

Fig. 7.1 Carissa Plains PV power station (Courtesy of P. D. Maycock, Energy Systems Inc.)

7.1.2 Applications of large PV projects

Large PV projects have been built to provide electricity in many different applications (Table 7.1). It should be borne in mind that many of these applications were demonstration projects, built to study the feasibility as well as the technical and economic potentials of this new technology.

The experience from these applications has demonstrated that, unlike small-scale applications, large PV projects are less cost-effective as stand-alone than in grid-connected operation. This is due to the fact that grid-connected config-urations can utilise all the produced PV energy, thus reducing the unit cost of generated electricity. Stand-alone applications, on the other hand, are heavily conditioned by the actual power demand of the isolated user which often means that the excess power cannot be utilised.

Table 7.1 Applications of large PV projects

- Grid-connected power production
- Stand-alone electrification of villages or communities
- Switchable electrification allowing both isolated stand-alone and grid-connected operation
- Water desalination and treatment (stand-alone)
- Water pumping (stand-alone)
- Cold-storage for food conservation (stand-alone)

7.1.3 Economics of large PV projects

There is little doubt that, at present, large PV projects are generally not competitive with the conventional power sources. The principal reasons are as follows.

- Investment costs are much higher, leading to much higher unit cost of generated electricity (see Table 7.2).

- Large PV power plants present technical complexity similar to the conventional power plants, requiring skilled personnel for their management and maintenance. Unlike small-scale PV systems they are therefore inappropriate for application in less-developed remote areas as they cannot provide the simple and reliable power source that is required in these locations.

- Conventional power facilities take advantage of the economy of scale, i.e. the larger the facility the lower the specific production cost (the cost of a unit of generated electricity). Large PV plants, on the other hand, do not present similar advantages since, under given climatic conditions, the required active surface of PV modules, support structures and other facilities is simply proportional to the energy output of the plant.

Table 7.2 Cost of electricity generation

Technology	Investment costs		Unit cost (US$/kWh)
Large thermal power plant	1.5	US$/W	0.06
Diesel power plant	0.5–1.0	US$/W	0.2–0.5
Large PV power plant	12–15	US$/W$_p$	0.8–1.0

For these reasons, the industrial applications of large PV projects have been, with a few exceptions, limited to industrialised countries where government or transnational programmes have been able to promote the development of PV technology specifically in large-scale applications (mostly grid-connected power plants).

7.1.4 Technical characteristics

Large PV plants are generally characterised by the following features:

- high-voltage DC bus (typically 200–600 V_{DC}),
- large-scale PV plants are generally AC systems which employ one or more inverters for the conversion of the DC power generated by the PV array to standard AC power.

The different types of large PV systems are shown schematically in Fig. 7.2. In common with other PV systems, they can be utility grid-connected or

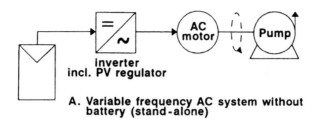

A. Variable frequency AC system without battery (stand-alone)

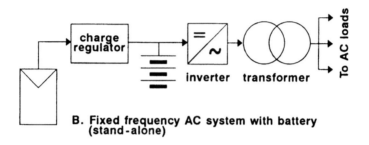

B. Fixed frequency AC system with battery (stand-alone)

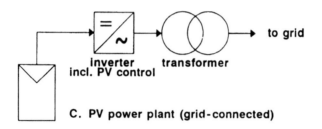

C. PV power plant (grid-connected)

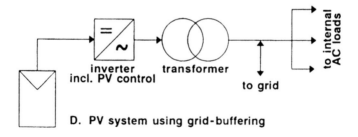

D. PV system using grid-buffering

Fig. 7.2 Different types of large PV systems

stand-alone. Stand-alone systems may, or may not, include a battery. The installation of battery storage in grid-connected systems gives little advantage since the grid may be used as a much more effective, unlimited and cheaper energy storage.

7.1.4.1 Stand-alone PV systems

Variable-frequency AC systems without battery are used where the application requires exclusively drive/shaft-power and an external energy storage is available. A typical application is for water pumping into a reservoir. These systems are characterized by a special variable-frequency inverter powering an AC motor. The drive-motor speed and, consequently, the power demand of the user is adapted instantaneously to the available sun power and makes the maximum use of the available PV power. The main advantages of such systems are high efficiency, rugged design and the possibility to feed any off-the-shelf standard AC motor which can operate at a variable speed (typically pumps, fans, etc.).

AC systems with battery represent the standard configuration of a PV power plant supplying an isolated (island) grid and may serve to electrify a community in a remote area. In this case one or (preferably) more fixed-frequency inverters feed continuously (24 h per day) standard AC power into a distribution grid.

Hybrid systems. The cost-effectiveness of a stand-alone PV power plant can often be improved by taking advantage of a hybrid (PV/wind or PV/diesel) design. During prolonged low-irradiation (no-sun) periods, the diesel genset (or other power source) then allows either to charge the main battery of the system, or to supply directly the AC users.

7.1.4.2 Utility grid-connected systems

PV power plants feed PV power directly and instantaneously into a utility grid by means of one or more inverters and transformers. In most larger systems, line-commutated inverters are used for such purpose, equipped with a maximum-power-point tracker.

PV systems using grid-buffering use the grid as energy storage (instead of a battery) to integrate the power demand of the load. During the day, excess PV power is fed into the grid while during the night the power required by the load is reclaimed from the grid. Obviously the user of such a system may minimise the energy exchange with the grid (and relevant costs) by adapting consumptions to the available sunpower.

7.1.5 Subsystems

The structure of PV systems and their subsystems were discussed in detail in Chapter 4. Here we focus attention on various features that are specific to large PV systems.

7.1.5.1 PV generator

Selection of the DC bus voltage. In general terms one can say that the optimum performance favours high voltage giving higher conversion efficiencies, and requiring smaller conduit sections and smaller (and less expensive) equipment.

However, the DC bus voltage cannot be increased in an unlimited way for the following reasons.

- *PV modules* are designed for low-voltage applications and do not present sufficient insulation distances between live circuit parts (solar cells) and the grounded metal frame as required for high-voltage applications.
- *Safety regulations* define voltage thresholds to distinguish between different voltage classes (low/medium/high) and relevant requirements to be satisfied by equipment in these voltage ranges.
- Present *inverter technology* accepts only limited DC voltage levels depending on the type of power-switching device adopted. Presently only SCR thyristors allow DC voltage levels exceeding $1000\,V_{DC}$.
- Standard *switching devices* for DC voltages above $500\,V_{DC}$ are difficult to find since commercial equipment is generally designed for AC application and not to assure interruption of the opening arc occurring in high-voltage DC applications.

7.1.5.2 PV array lay-out

As we have seen in Chapter 4, the power rating of the PV array and the DC bus voltage specification define directly the electrical series/parallel configuration of the PV array.

In some instances, it may be advantageous to subdivide a large PV generator into electrically separate subarrays, particularly in the following situations.

- Different types of PV modules are to be installed in the same plant.
- Different array sections have to be installed at different orientations.
- The application requires the possibility to partialise (section) the PV generator.

Once the electrical array configuration is defined, the physical layout of the PV array in terms of array sections (corresponding to electrical subarrays) and rows (containing single or multiple strings) may be determined.

7.1.5.3 Utility grid interface of PV power plants

Large-scale PV power plants are generally connected to a medium-voltage utility grid section of typically 10–$20\,kV_{AC}$. The components of the grid interface which are generally included for this purpose are summarised in Table 7.3.

7.1.5.4 System control

In the early days of automatic control, all control functions of an installation were usually concentrated in one powerful (and expensive) central computer. Such a solution was believed to be cheaper and simpler since powerful software was expected to do all the work.

Table 7.3 Grid-interface components

- Step-up transformer(s)
- Power factor correction (capacitors) if required (e.g. for SCR inverter)
- Filtering for power quality improvement (harmonics)
- Grid interface protections which monitor:
  - maximum current limit
  - voltage range (max/min thresholds)
  - frequency range (max/min thresholds)
- Grid interface instrumentation, including at least:
  - active energy counters (separate for in/outgoing)
  - reactive energy counters (separate for in/outgoing)

Frequently, an entire installation became dependent on the correct operation of the process controller, a very sensitive component of low reliability. As a result, nobody was able to operate the plant manually without the computer. And even worse, if the computer failed, there was no way to find out what had happened.

Modern control strategies therefore tend to adopt distributed controls: all control functions are split up into single and independent functional blocks or loops which are implemented by front-end control devices installed inside or nearby the addressed equipment. The coordination between different units is assured via shared process parameters (for example, battery voltage or PV bus voltage). The functions of the central automatic process controller (when included) should be limited to the minimum, i.e. to supervise and coordinate operation of the overall plant.

The control system of a large PV plant should include the following *four levels of control*.

1. *Process instrumentation* measures the status and values of process parameters and transmits this information to a higher-level controller (low-level control device, process controller or data acquisition system, or an indicator for the human operator).
2. *Front-end (low-level) control devices* perform specific control functions (loops) not requiring high-level coordination with other plant components (inverter internal controls, power factor correction, etc.).
3. The high-level *'intelligent' process supervisor* coordinates and supervises the operation of the entire plant, and acts as an interface with the human operator.
4. The *human operator* must be able to obtain easily all required information to provide the highest-level supervision of the plant.

Data acquisition (logger) functions may also be executed by the automatic plant supervisor, since the hardware is usually capable of performing this role. However, a separate unit specifically tailored for this purpose is usually more appropriate for the maximum reliability of data monitoring.

7.1.6 Safety aspects of high-voltage PV systems

As a result of the high DC bus voltage, large PV systems require specific safety measures which include personnel protection and grounding of all metal parts, including PV module frames. The safety measures must bear in mind the principal consideration that *the PV generator cannot be switched off effectively during daylight*.

Table 7.4 gives a brief outline of the main protection measures that PV power stations should be provided with. Some of these measures have already been discussed in other parts of this book but protection with respect to lightning strikes should be particularly noted. Experience shows that lightning damage is usually suffered by the sensitive power conditioning and control electronics, rather than by the PV modules themselves. The lightning is most likely to strike the support structures of the PV array which are generally made of galvanized steel and well grounded. The utility grid also suffers frequent lightning strikes and conveys part of the resulting surges through the grid interface into the PV power plant.

Summary

The special features of large PV systems have been reviewed, and their applications and economics contrasted with smaller applications and conventional power sources. The technical characteristics were discussed with focus on the subsystems. It was shown that large PV generators are usually characterised by high DC bus voltage, leading to specific safety and protection measures. Other components of large PV systems, the grid interface and system control, have also been examined.

Table 7.4 PV plant protection

- *DC shorts:* blocking diodes installed on all parallel strings
- *AC shorts* to be counteracted up-stream (inverter side) and down-stream (grid interface)
- *Grid interface protection:*
  - PV plant from the grid
  - Grid from the PV plant
- *Monitoring of insulation failures* which are relatively frequent in PV modules
- *Lightning and surge protection:*
  - Effective grounding of all metal structures
  - Galvanic separation of circuits in different areas of the plant
  - Overvoltage (surge) protection
  - Grounded Faraday-cage shielding of sensitive control electronics
- *Fencing* to avoid theft and vandalism

7.2 PHOTOVOLTAICS UNDER CONCENTRATED SUNLIGHT

7.2.1 Introduction

Concentration of sunlight is easily demonstrated by holding up a magnifying glass to the sun. The bright focus of light formed is intense enough to set fire to paper or wood. Place a photovoltaic cell at this focus, and you have a concentration photovoltaic system. Equally well, the concentrated energy could be used to drive a heat engine, as is the case with solar thermal systems.

In the case of photovoltaics, why should we be interested in using a concentrator? The main reason is to reduce the cost of the cell by making it smaller, whilst we still collect light from a large area (Fig. 7.3). For this to make sense, the concentrator has to be cheaper than photovoltaic cells of the same area. Another reason is that, once we are producing electricity from small cells, it becomes attractive to use more sophisticated, higher-efficiency types of cell which would be too expensive to use without concentrators. Such is the case with gallium arsenide cells, for example. Some of the most exciting developments currently taking place in solar-cell research may find their first use in concentration systems.

Cells designed to work in concentration systems produce a higher efficiency than is possible with unconcentrated sunlight. In Fig. 7.4 we see how the record efficiencies for cells with and without concentration have risen since 1978. The highest efficiencies over 30%, achieved since 1988, are a result of using two types of cell stacked on top of each other, thus making good use of a broader part of the solar spectrum (see Chapter 3). So far 34% efficiency has been obtained in this way, using 100 times the concentration of sunlight. With three cell types and concentration, researchers hope to raise the figure to 40% in the next few years.

The efficiencies for cell and concentrator together are lower, due to loss of light in the concentrator and to cell heating, which both decrease efficiency. The record for prototype photovoltaic modules is therefore only about 23%.

Certain drawbacks have so far prevented concentration systems from making a significant penetration into the world photovoltaic market.

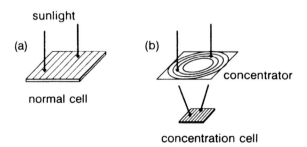

Fig. 7.3 (a) A normal solar cell in unconcentrated sunlight. The concentration cell (b) is smaller and cheaper but the concentrator has to be paid for

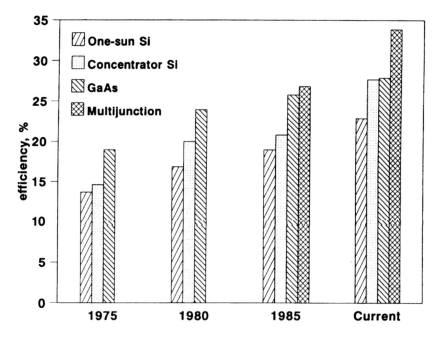

Fig. 7.4 Record efficiencies with and without concentration: silicon, and concentration GaAs and multijunction cells (after E. C. Boes and A. Luque, in T. B. Johannson, H. Kelly, A. K. N. Reddy and R. H. William, *Renewable Energy — Sources for Fuels and Electricity*, Earchscan: Island Press, London/Washington DC., 1993)

- Concentrators only use the direct beam light but not the diffuse light scattered by clouds and the atmosphere. This means that they are only useful where there is a large amount of direct sunlight. Even for a cloudless sky, the diffuse light can constitute 15–30% of the total solar radiation available.
- Concentrators always have to be pointed towards the sun. The mechanical system producing the required motion is called a *sun tracker*. The additional complexity of the tracker has so far disfavoured use in remote locations because of reduced reliability and increased maintenance needs.

These two drawbacks do not apply to certain types of *static concentrators*, but these are not capable of concentrating sunlight more than about 10 times, whereas tracking concentrators can achieve much higher concentrations as we shall see.

Given these considerations, concentration systems are most likely to have significant advantages over conventional systems for central power generation (in sunny locations) for national or local grids (sections 5.7 and 7.1). Here, greater complexity can be justified and the constraints on cost and efficiency are most stringent.

The two largest concentration photovoltaic systems to have been built to date use Fresnel lenses and each have a peak output of about 300 kW. The first was completed in Saudi Arabia in 1981 and has an efficiency of about 9%. The second was installed in Texas, USA, in 1989, on the roof of a garage. Its efficiency is about 13%. These efficiencies are calculated from the direct current output of the system, and relative to the direct component of sunlight.

A number of other systems of output power greater than 10 kW have also been constructed. The power output of these installations is small compared to that of solar-thermal stations, the largest of which produces up to 100 MW. But perhaps this needs to be set against the fact that photovoltaics has only been a subject of serious research since the 1950s, whereas the heat engines used in solar-thermal installations have been under development for over two centuries.

7.2.2 Concentration and efficiency

Before studying optical systems which can be used to concentrate sunlight, we shall look at how solar cells behave under concentration.

In Chapter 3 we have seen that the short-circuit current I_{sc} is proportional to the irradiance of the cell. Under concentration, therefore,

$$I_{sc}(X) = XI_{sc}(1) \tag{7.1}$$

where we denoted explicitly the dependence on the *optical concentration ratio*, X. The open-circuit voltage, on the other hand, is given by

$$V_{oc}(X) = (kT/q)\ln\left(\frac{I_{sc}(X)}{I_o} + 1\right) \approx (kT/q)\ln\left(X\frac{I_{sc}}{I_o}\right)$$
$$= V_{oc}(1) + (kT/q)\ln X \tag{7.2}$$

where I_o is the diode dark saturation current, q is the magnitude of the electron charge, T is the temperature and k is the Boltzmann constant. Therefore, noting that the irradiance $G(X)$ increases in proportion to X, the efficiency as a function of X is:

$$\eta(X) = \frac{V_{oc}(X)I_{sc}(X)FF}{G(X)} = \eta(1)\left(1 + \frac{(kT/q)\ln X}{V_{oc}(1)}\right) \tag{7.3}$$

The relative increase of efficiency of the ideal solar cell is therefore proportional to the natural logarithm of the concentration ratio. For example, concentrating light tenfold ($X = 10$) gives the voltage increase $(kT/q) \ln X$ equal to $0.026 \times 2.3 = 0.06$ V at 25 °C. Since the open-circuit voltage of a silicon cell is typically 0.6 V without concentration, this amounts to about a 10% increase.

The efficiency of practical devices, however, cannot increase indefinitely. As we have seen in Chapter 3, a solar cell has a finite series resistance R_s. Power dissipated as heat in the series resistance is:

$$P_{loss} = I^2 R_s. \tag{7.4}$$

The current I flowing from the cell is proportional to the concentration X. On the basis of equation (7.1), the power wasted is:

$$P_{loss} \cong X^2 I_{sc}(1)^2 R_s \qquad (7.5)$$

which grows rapidly as X is increased. It can be shown (Luque, 1989) that the optimum concentration is about:

$$X \cong (kT/q)/I_{sc}(1)R_s \qquad (7.6)$$

For example, suppose a cell produces a current of 50 mA at 1 sun ($1\,kW/m^2$) and has a series resistance of $0.01\,\Omega$. With $kT/q = 0.026\,V$ (at 25 °C), we can estimate the optimum concentration ratio X to be about 50, corresponding to an irradiance of $50\,kW/m^2$.

Although cells designed for concentration incorporate special features to reduce the series resistance, it is difficult to make a silicon cell with good efficiency at values of X much greater than about 200 (i.e. $200\,kW/m^2$).

7.2.3 Maximum possible concentration

The maximum concentration ratio that can be achieved is affected by three principal factors.

- On a fundamental level, one must make allowance for the fact that the Sun's rays that arrive at the Earth are not exactly parallel.
- If the cell is adjoined to a refractive medium, such as glass, this can make higher concentrations possible.
- In practical situations, the concentration ratio is also limited by tracking errors of the system.

These factors give the following limit on the concentration ratio:

$$X_{max} = n^2/\sin^2\theta_c \qquad (7.7)$$

where $\theta_c = \theta_{sun} + \theta_e$, θ_{sun} is the half-angle subtended by the Sun's rays, θ_e is the tracking error (typically in the range $0.1°$ to $3°$), and n is the refractive index of the adjoining medium (n is not usually greater than 1.6 for economic concentrators). The principle behind this formula is that sunlight cannot be concentrated beyond the point where it could heat an object to a temperature higher than that at the surface of the sun itself (Winston, 1991).

Since $\theta_{sun} = 0.27°$, the limit in the case of $n = 1$ and $\theta_e = 0$ (i.e. a solar cell surrounded by air in a system with no tracking errors) is $X_{max} = 45\,000$. The record concentration to have been achieved experimentally is $84\,000$, using sapphire for which $n = 1.76$ (Cooke et al., 1990). The theoretical limit for this refractive index is $140\,000$.

Concentrators which attempt to reach very high concentrations (> 200) usually have rotational symmetry, like those in Fig. 7.5. Concentrators of linear symmetry, such as those shown in Fig. 7.6, are often simpler to construct because they can be made by bending flat materials, but they achieve lower concentrations. The limit in their case is:

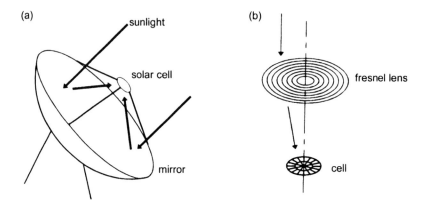

Fig. 7.5 Point-focus systems with rotational symmetry, (a) Dished parabolic mirror; (b) Fresnel lens

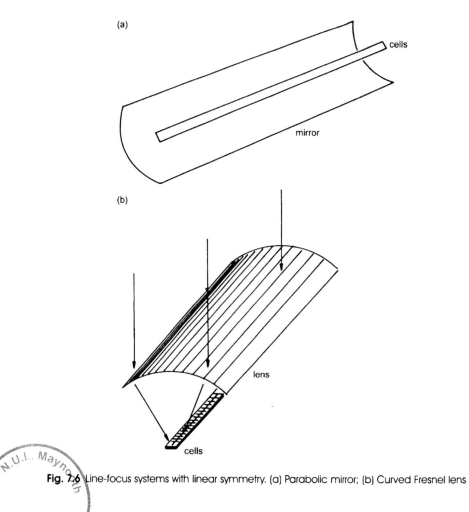

Fig. 7.6 Line-focus systems with linear symmetry. (a) Parabolic mirror; (b) Curved Fresnel lens

$$X_{max} = n/\sin\theta_c \qquad (7.8)$$

which is only the square root of X_{max} as given by equation (7.7).

7.2.4 Design of practical concentrators

We shall now look at some real concentrators to see how they compare with the ideal maxima of equations (7.7) and (7.8). We often find that the concentrations obtained are much lower. However, this is not a great problem because the ideal limits are, to begin with, usually higher than the optimum of equation (7.6).

The concentration ratios we present, in terms of the collection angle θ_c, are those which could be obtained if there were no losses of light in the concentrator. In practice losses amount to at least 10%, due to absorption and Fresnel reflections, depending to a large extent on the quality of the materials used.

7.2.4.1 Focusing systems

Focusing systems use parabolic lenses or convex mirrors to focus parallel rays on a spot or line, as shown in Figs 7.5 and 7.6. Because, however, the Sun's rays are not absolutely parallel, the concentrated light does not come to a sharp focus but falls within a small area. Also, tracking inaccuracy can cause this illuminated area to be shifted off centre. The cell must be big enough to capture the illuminated area in all its possible positions, and this reduces the concentration ratio. However, a very large tracking error is always unacceptable for these systems because, even if they are designed to take it into account, the result is that just a small part of the cell is illuminated. Such gross non-uniformity of illumination decreases considerably the efficiency—just like a non-uniformly illuminated flat panel, as we saw in Chapter 4.

Let us now look at a practical example. For a dished parabolic mirror, like that of Fig. 7.5(a), with the focal length of the mirror properly chosen, the maximum concentration obtainable is, for small collection angles (Welford and Winston, 1989)

$$X_{max} = 1/(4\theta_c^2) \qquad (7.9)$$

which is just a quarter of the maximum given by equation (7.7). Figure 7.7(a) shows the variation of concentration with collection angle θ_c for the dished parabolic mirror, in comparison with the ideal limit; in Fig. 7.7(b) we show the situation for the line-focus system, where the concentration is just the square root of that given by equation (7.9).

The concentrations possible with mirrors are quite adequate for use with solar cells. In the case of the dished mirror (i.e. rotational symmetry), they tend to be higher than desired, although this is not an obstacle because the concentration can always be reduced by moving the cell closer to or further away from the mirror.

However, mirrors have some important drawbacks. Because the cell is mounted in the sunlight, it tends to be more difficult to cool it. The cell and cooling fins cast a shadow on the mirror, which decreases the power output of the system.

These factors have tended to favour Fresnel lenses over mirrors. The advantage of Fresnel lenses over solid lenses is that they are thinner and so use less material and are cheaper.

The equations for the maximum concentration are more complicated for lenses than mirrors. Here we shall only look at the results, which are summarised in Fig. 7.7(a) for point focus and Fig. 7.7(b) for line focus.

Point-focus Fresnel lenses (Fig. 7.5(b)) usually have a smooth, flat upper surface because this is easiest to fabricate and clean. Line-focus lenses (Fig. 7.6(b)) are easy to curve, which results in advantages in the concentration which may be obtained, as compared to flat line-focus lenses, although the concentration is always less than that obtained for point-focus systems.

The concentration obtained from lenses is lower than from mirrors, but still high enough to make practical systems. Indeed several such systems have been constructed. Figure 7.8 shows a line-focus concentrator, like that in Fig. 7.6(b), in which the lens and the side walls which connect the lens with the cells are made as a one-piece extrusion. This lowers the cost of the module, and provides a good protection for the cell against the environment.

7.2.4.2 Non-imaging concentrators

The lenses and mirrors discussed above are similar in principle to those used in optical instruments such as telescopes, microscopes, photographic cameras, etc. All these instruments are used to produce an image. Similarly, focusing concentrators produce an image of the sun on the solar cell. The production of an image is, however, superfluous in the case of solar concentrators. All that matters is that the Sun's rays should reach the cell. This approach leads to designs called *non-imaging concentrators*.

The most important type of non-imaging concentrator is called the compound parabolic concentrator (CPC), shown in Fig. 7.9(a). It consists of two parabolic surfaces each having its focus at opposite extremes C and C' of the cell. This means that rays entering the concentrator at an angle of $+\theta_c$ or $-\theta_c$ (measured to the axis of the concentrator) are reflected towards C and C' respectively. By so designing the concentrator, we make sure that rays entering within a smaller angle reach the cell at some point between C and C'.

The concentrations reached by trough-shaped CPCs (i.e. those of linear symmetry as shown in Fig. 7.9(b)) are equal to the absolute limit given by equation (7.8). For CPCs of rotational symmetry (Fig. 7.9(c)), the concentrations are just slightly below the absolute limit (see Fig. 7.7(a)).

The possibility of using high acceptance angles and still achieving reasonable concentrations makes the trough CPC attractive for *static concentrators*. These do not use any tracking system, but make use of the fact that the elevation angle of the sun above a plane parallel with the equator is restricted to $\pm 23°$ about

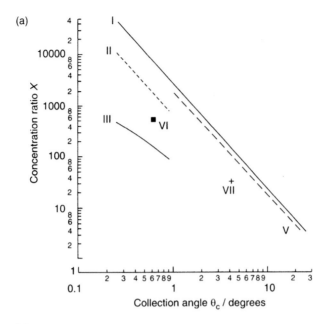

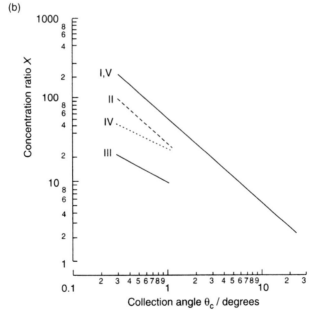

Fig. 7.7 Concentration ratio X as a function of the collection angle θ_C for concentrator with (a) rotational and (b) linear symmetry. I: Theoretical limit for $n = 1$ (equations (7.7) and (7.8)). II: Limit for parabolic mirrors. III: Limit for flat Fresnel lenses. IV: Limit for shaped Fresnel lenses (not shown in (a)). V: Limit for compound parabolic concentrators. VI: Example of a two-stage system using non-imaging secondary concentrator (Fig. 7.10(a)). VII: Example of a two-stage system using imaging secondary concentrator (Fig. 7.10(b))

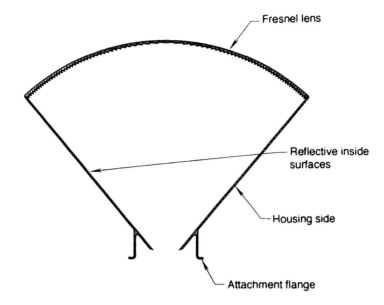

Fig. 7.8 Cross section through a curved, linear Fresnel lens. The lens and housing walls are extruded in one piece (courtesy of SEA Corporation)

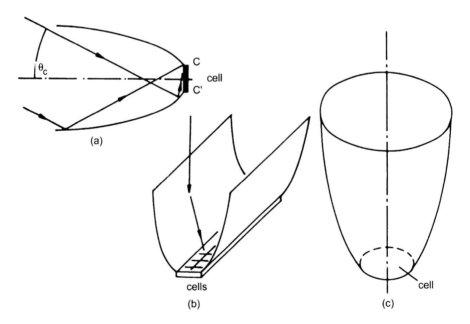

Fig. 7.9 (a) Cross section through CPC showing how rays at an angle θ_C to the axis of symmetry are directed to one extreme C of the solar cell. Similarly, rays entering at $-\theta_C$ are directed towards C'. (b) Trough-shaped CPC of linear symmetry. (C) Funnel-shaped CPC of rotational symmetry

its mean value throughout the year. We can therefore accept this variation as a tracking inaccuracy. The result is that a concentration of about 2.5, or greater if the cell is adjoined to a medium of refractive index higher than unity, can be obtained. Larger concentrations are possible if small adjustments are made manually to the orientation of the concentrator throughout the year. This might be attractive in locations where technological resources are limited.

A drawback of CPCs is that, for high concentration, they have to be very long in relation to the diameter of their entrance aperture. This makes them unwieldy and expensive for concentration ratios above about 50. Another disadvantage of the CPC is that it does not guarantee uniform illumination of the solar cell.

7.2.4.3 Two-stage concentrators

Almost all of the disadvantages of focusing and non-imaging concentrators are overcome by using concentrators having two stages, called a primary and secondary concentrator. The relation between concentration and acceptance angle can be close to the ideal limit, the problem of non-uniformity of illumination can be corrected, and the concentrator can be quite compact.

The only disadvantages are the extra complexity and cost, and the additional optical losses which can occur in the second stage.

A focusing optical element, such as a flat Fresnel lens, is generally used for the primary concentrator. The secondary can either be a non-imaging mirror (see Fig. 7.10(a)), similar in principle to the CPC of Fig. 7.9(c), or an imaging lens (see Fig. 7.10(b)). The non-imaging type gives very good tolerance to tracking errors, whereas the imaging secondary is particularly attractive in allowing a very uniform illumination of the cell.

Because so many different designs of two-stage concentrator are possible, it is not easy to summarise their performance. However, we show in Fig. 7.7(a) the concentrations and collection angles for the examples of Figs 7.10(a) and 7.10(b). Compared to the CPCs' achieving equivalent concentrations, their length is reduced 3 to 9 times.

7.2.5 Tracking systems

We have seen how (apart from the static concentrators) tracking systems are needed in concentration systems, and the more accurate the tracking, the greater the concentration which can be obtained. In Chapter 2, the motion of the Earth around the Sun and its effect on sunlight have been described. What the tracker has to do is move the concentrator modules in such a way as to cancel out this motion, keeping their orientation relative to the Sun fixed.

In the polar-axis system, for instance, the main axis of rotation is parallel to the Earth's polar axis. The rotation about this axis is at the same speed as the Earth's rotation, but in the opposite direction. Then, throughout the course of the year, it is necessary for the panels to be tilted relative to the polar axis because of the Earth's declination.

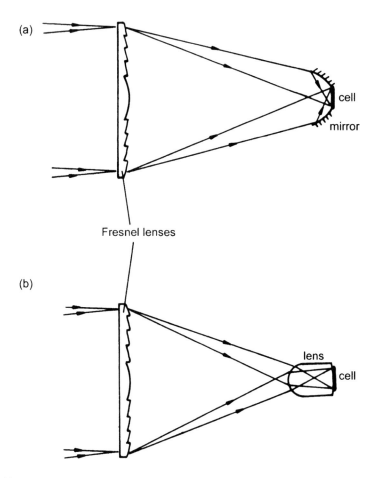

Fig. 7.10 Two-stage concentrators in which the primary stage is a Fresnel lens. (a) Non-imaging secondary stage similar to CPC. (b) Imaging secondary stage consisting of a single-surface lens

A practical example of a polar-axis system is shown in Fig. 7.11. An interesting feature of this design is that, unlike most trackers which are driven by electric motors, it is worked by fluids heated directly by the Sun. The heat causes the fluids to expand and move pistons.

A different type of tracking mechanism is the azimuth-elevation type, an example of which we see in Fig. 7.12(a). The azimuth axis is vertical. At sunrise the concentrator modules face east, then during the day rotation around the azimuth axis turns them round to face west at sunset. At the same time, the elevation axis (horizontal and mounted on top of the azimuth axis) tilts the module upward as the Sun rises up till noon, then down again as dusk approaches. An interesting feature of this system is the way the modules are staggered, leaving spaces for the wind to pass through. This reduces the wind loading, and thus the strength required of the structure. Figure 7.12(b) shows the module used in this machine.

Fig. 7.11 Polar axis tracking system being used to pump water in Bangladesh. The heat of the sun is used to produce the tracking motion via a mechanical system (courtesy of Midway Labs, Inc.)

Although the motions of the azimuth-elevation tracker are more complicated and difficult to control than those of the polar-axis type, an advantage seems to be that the structure does not have to be so high (relative to the module size) to avoid the modules hitting the ground.

The above examples both use two axes of motion. One-axis systems are possible with non-imaging concentrators and some line-focus systems. In Fig. 7.13 we give an example, a venetian-blind type tracking system, used with the linear Fresnel lenses as of the kind shown in Fig. 7.8.

Summary

Concentration can be advantageous because of the reduction in the area of photovoltaic cells it achieves, but it introduces extra complexity in terms of the need to track the Sun. The output voltage and hence efficiency of cells tend to increase under concentration but, under very high concentrations, losses through the electrical resistance of the cell cause the efficiency to decline.

There is a theoretical limit to how many times light can be concentrated. The limit is lowered by tracking inaccuracy which is significant in all practical systems. Dished parabolic mirrors only achieve 1/4 of the theoretical limit

(a)

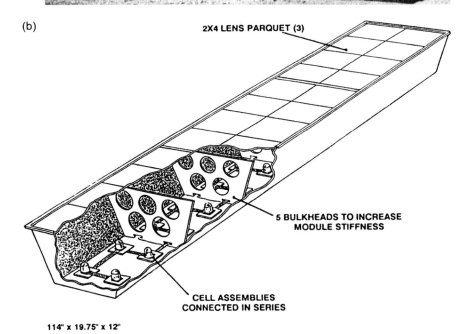

(b) **2X4 LENS PARQUET (3)**

**5 BULKHEADS TO INCREASE
MODULE STIFFNESS**

**CELL ASSEMBLIES
CONNECTED IN SERIES**

114" x 19.75" x 12"

Fig. 7.12 A large azimuth-elevation tracking system designed to give 15 kW power output. (a) The modules are staggered to leave gaps for the wind to pass through which reduces the load on the structure. (b) Cross section through the module used in this system which is similar in principle to that in Fig. 7.10(b) (both figures courtesy of Alpha Solarco)

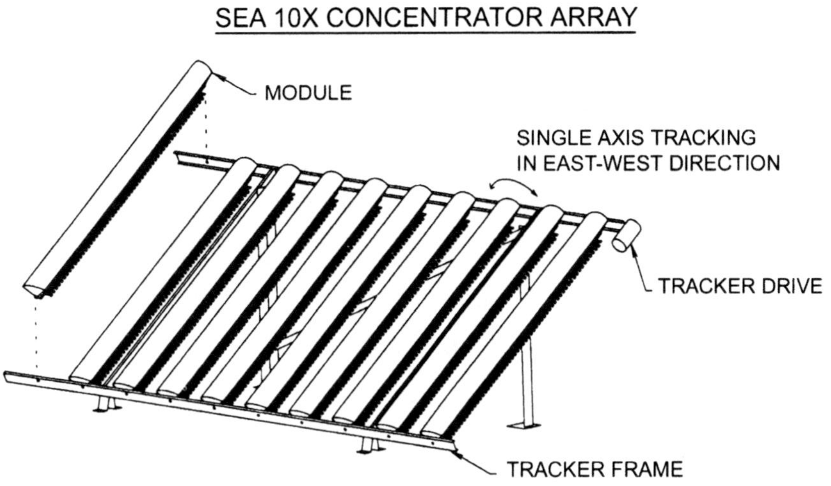

SEA 10X CONCENTRATOR ARRAY

MODULE

SINGLE AXIS TRACKING
IN EAST-WEST DIRECTION

TRACKER DRIVE

TRACKER FRAME

Fig. 7.13 Venetian-blind type one-axis tracking system for line-focus Fresnel lenses of the kind shown in Figs 7.6(b) and 7.8 (Courtesy of SEA Corporation)

and lenses achieve less. But lenses are still adequate for practical systems provided the tracking accuracy is good, i.e. better than $\pm 1°$. Non-imaging concentrators such as the compound parabolic concentrator permit greater inaccuracy, but are unsuitable for large concentrations (> 50) because of their excessive size. Two-stage concentrators combine (at the expense of greater complexity) high concentration, tolerance to tracking inaccuracy, and small size.

Various tracking mechanisms have been built. The one-axis systems are simpler than the two-axis systems, but cannot be used to obtain such high concentrations.

7.3 ELECTROCHEMICAL PHOTOVOLTAICS

7.3.1 Photoelectrochemistry presented

Photovoltaics is usually understood to imply a technology developed around solid-state semiconductor devices. It is therefore instructive to recall that when Alexandre-Edmund Becquerel (Becquerel, 1839) first observed a photoelectric effect, it was at the interface of a semiconductor with a liquid. The semiconductor in question was silver chloride, in which there was then much interest because of the contemporary emergence of photography, with the recognition of the work of Daguerre and the presentation to the Royal Society in London of the halide process by Fox-Talbot. Becquerel's observation was the first evidence that photography exploits an electrical charge transfer effect. The parallel development of photography and photoelectrochemistry will re-emerge in this chapter when dye sensitisation is presented. In the Becquerel experiment, the asymmetrical behaviour of two identical immersed electrodes was noted when one of them was illuminated. The generation of a photovoltage when light is incident on a solid-state junction (metal-semiconductor junction in the Schottky device or the homo- or heterojunction between two semiconductors) was discussed in Chapter 3. By analogy, the Becquerel effect is observed when a semiconductor is in contact with an electrolyte where the carriers are ions.

7.3.2 Electrochemical photovoltaics

7.3.2.1 Regenerative cells

A redox couple is a mixture of the oxidised and reduced forms of the electro-active species in a redox system. The equilibrium potential of a redox couple, as measured by a nonpolarised electrode in contact with it, can be regarded as the equivalent of a Fermi level. By analogy with the discussion of p-n junctions in Chapter 3, any semiconductor in contact with that redox electrolyte in darkness must arrive at a common equilibrium potential by charge exchange. Assuming that the redox component concentration will typically be of molar order (in other words, Avogadro's number of ions per litre) the charge carrier density will be some 10^{20} cm^{-3}, a high multiple of that in the semiconductor. The effect on the semiconductor will therefore be similar to the Schottky junction with a metal. As a consequence, a layer with depletion of majority carriers forms near the interface, with an associated space charge and electric field. The near-interface zone is therefore positively charged in an n-type semiconductor, with an upward deflection of the band edges, and inversely for a p-type material. When charge carrier pairs are photogenerated by light absorption in the semiconductor, the minority carriers migrate to the interface where, if not lost by recombination, they are available for transfer to the electrolyte, and can engage in an electrochemical reaction. At the surface of an n-type semiconductor under illumination, therefore, there is a population of positively charged holes capable of initiating an oxidation. This type of semiconductor is thus a photoanode. Photocathodic effects take place at p-type semiconductor interfaces with electrolytes. The majority carriers are displaced towards the bulk of the semiconductor, where they can be collected by an ohmic contact as in a conventional solid-state PV device. In the n-type material the negative charge of the electrons displaced in this fashion raises the Fermi level, representing a photovoltage generated by the device, so that the ohmic back contact is negative with respect to the electrolyte interface, and therefore in the case of a regenerative cell, with respect to the redox level. This photovoltage therefore provides the electromotive force for the current in the electrical load, completing the circuit through the counter-electrode to the electrolyte (Fig. 7.14). With a redox electrolyte, as already mentioned, the effect of the electrons at a metal counter-electrode acting as a cathode will complement the oxidation at an n-type semiconductor photoanode, maintaining the electrolyte composition constant. In this ideal case, therefore, the conversion of light to electricity takes place as in a solid-state junction device. Considerations of open-circuit voltage, current dependent on light intensity, fill factor and recombination losses apply as in the conventional device.

7.3.2.2 Photoelectrolytic cells

It will be noticed in Fig. 7.15 that several semiconductors have a conduction band edge more negative than the potential required for hydrogen evolution and at the same time a valence band at a level compatible with hole transfer to the electrochemical reaction of oxygen evolution. Materials with this combination of properties (bandgap greater than the difference of hydrogen and oxygen evolution reactions, and an appropriate Fermi level position relative to NHE when immersed in aqueous electrolyte) are, in principle, capable of photoelectrolysis of water. However, as noted in Chapter 3, the effective photovoltage generated by a junction is significantly less than the band gap of the semiconductor. Silicon with a bandgap of 1.12 eV, for example, gives a typical photovoltage of 0.6 V. Taking account also of over-potential requirements, a practical minimum bandgap of about 2.4 eV must be specified for photoelectrolysis. In fact, the first reported photoelectrolysis experiment (Fujishima and Honda, 1971) used an n-type titanate photo-electrode with a bandgap greater than 3.0 eV, and therefore responsive only to ultraviolet light. Evidently, such photoelectrodes are poor absorbers and converters of solar radiation, and therefore marginally efficient as practical devices. The obvious alternative is to use a two-photon cell, with complementary n-type photoanode and p-type photocathode, each of much narrower bandgap. That solution, in turn, is confronted with a fundamental problem of narrower bandgap semiconductors in photoelectrochemistry – photocorrosion. The width of the bandgap is a measure of the thermodynamic stability of solids. The wider bandgap representing stronger chemical bonding between the component atoms. Consequently, in n-type material, when an oxidising photogenerated hole arrives at the interface with the electrolyte, the more probable reaction may be the oxidation of the semiconductor itself rather than the evolution of oxygen. These photocorrosion processes have been the decisive defect of some otherwise promising photoelectrodes, such as cadmium sulphide, which can be stabilised only in regenerative systems with fast low-overpotential redox electrolytes but are excluded for photoelectrolysis. As of now, therefore, efficient direct photoelectrolysis is no more than an elusive goal and practical developments in photoelectrochemistry have taken another direction.

7.3.2.3 Dye-sensitised photovoltaics

The consequences of the restricted spectral response of photochemically active semiconductors was evident in 19th century photography although, of course,

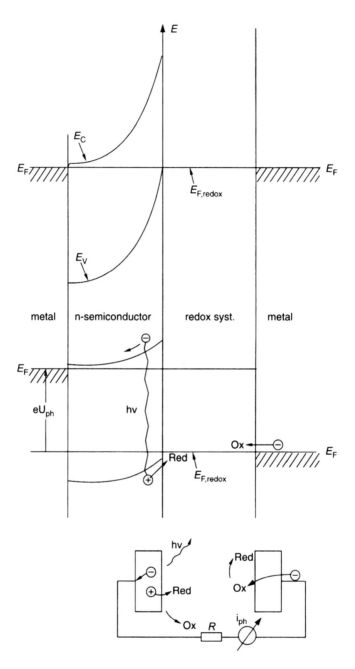

Fig. 7.14 Top: n-type semiconductor in darkness in contact with a redox electrolyte, using a metallic counter-electrode. Bottom: The same electrode under illumination, showing photovoltage generation (after Memming, 1988)

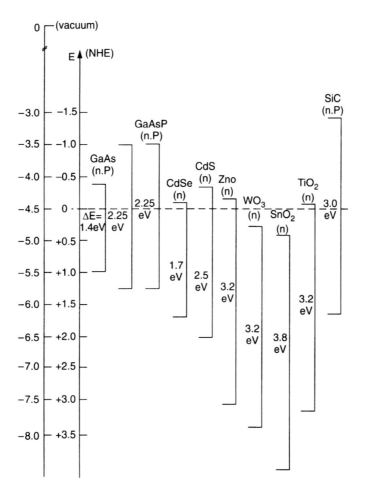

Fig. 7.15 Energy levels of valence and conduction band edges of some compound semiconductors in pH1 aqueous solution, on the physical scale relative to vacuum level and on the electrochemical scale relative to the reversible hydrogen electrode. Oxygen evolution in electrolysis occurs at a potential at least 1.23 V positive of NHE (after Memming, 1984)

there was then no technical explanation. In fact, the silver halides used in photoplates at that time had a bandgap of 2.7 to 3.2 eV and therefore responded only to blue and UV light. The consequent black-and-white rendering of scenes on 'orthochrome' photographic plates was somewhat unrealistic. Progress was empirical, with Vogel after 1873 (West, 1974) associating dyes with the gelatin and halide in a photographic emulsion to obtain 'panchromatic' sensitivity to the full visible spectrum, and eventually leading to the spectrally selective dyes in modern colour films. The electrochemical analogue, the sensitisation of a photoelectrode to longer wavelength light, followed shortly after (Moser, 1887).

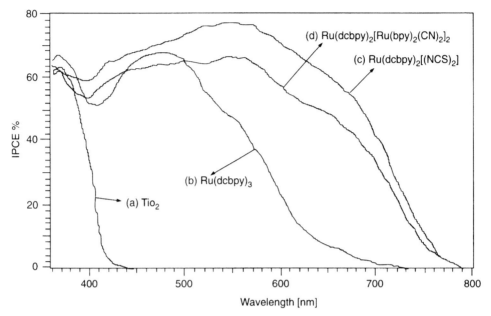

Fig. 7.16 Evolution of sensitising dyes, with increased incident photon conversion efficiency at longer wavelengths. (a) Titanium dioxide itself is transparent in the visible spectrum, absorbing only in the ultraviolet, wavelengths shorter than 400 nm. (b) When sensitised with the original dye (1989), a tris-bipyridyl carboxylate complex of ruthenium. (c) A cyanide-bridged trimer of dye (b). (d) Dye (b), spectrally modified by attached thiocyanide groups

The excitation and charge transfer mechanism was confirmed only in the 1960s (Gerischer and Tributsch, 1968), and is shown in Fig. 7.16. Since the semiconductor with its wider bandgap is insensitive to the incident light wavelength, the dye molecule becomes the light absorber, and is excited by photon capture to an energy level above the conduction band edge of the n-type substrate. It is therefore capable of injecting an electron into the semiconductor, becoming itself positively charged in the process. The uncharged ground state of the dye is recovered by reaction with a redox electrolyte. The injected electron migrates to the ohmic contact, passes through the external circuit to the counter-electrode, and regenerates the redox system.

Efficient photovoltaic devices based on that principle have since been developed (DeSilvestro *et al.*, 1985). A priority is the selection of a suitable dye, with a wideband absorption spectrum for visible light, and the excited level being appropriate for electron injection. The electroactive function of the dye requires that it be stable both in its ground, uncharged, state and in its oxidised state after electron loss to the semiconductor. Many candidate dyes are therefore transition metal complexes where the multiple permitted valence states of the metal allow the dye molecule to acquire or lose charge without disassociation. The dye must be intimately associated with the semiconductor surface for efficient charge transfer, an objective best achieved if it is chemically bonded to that substrate in a single molecular layer. However, since such thickness of dye absorbs light only weakly, the semiconductor should be rough and porous so that a sufficient optical density is achieved. Suitable dyes have been developed based on bipyridyl complexes of ruthenium, with substituted thiocyanide groups as spectral modifiers and carboxylate groups to chemically bond to the semiconductor substrate. The semiconductor, preferably titanium dioxide, is of course stable against photocorrosion due to its wide bandgap, conferring the added advantage that oxidising holes, which could otherwise attack the dye itself, cannot be generated under sunlight. The initial prototype described in 1991 had an AM1.5 conversion efficiency of 7.1%. Progress since then has been incremental, a synergy of improvements in structure, substrate roughness, dye photochemistry and electrolyte redox chemistry. That evolution continuing progressively since then, has now reached an independently-certified photovoltaic conversion efficiency of over 10%. The semiconductor structure, specifically, is now a nanocrystalline porous assembly, with a surface area enhancement factor, compared to the geometrical photoelectrode area, of about 10^3.

For the operating mechanism of the cell, it should be noted that with electron injection into the conduction band of an n-type material, we are here dealing with a majority-carrier device. As already noted, no holes are photogenerated and the oxidised species after charge separation is a charged molecule on the surface, not an electronic phenomenon in the solid phase. Consequently recombination losses as found in conventional junction devices are excluded. Therein is the explanation of the efficiency of the device despite its porosity and enhanced surface area, which would otherwise promote rapid surface recombination. Granted, a small proportion of the injected electrons can escape the solid by tunnelling, after diffusing against the space charge field, and may neutralise oxidised dye molecules, or more probably react reductively with a component of the redox electrolyte. To distinguish from solid-state recombination in a minority-carrier device, it is preferred to refer to these loss processes as sensitiser recapture and redox capture, respectively.

From a production and engineering aspect these dye-sensitised photoelectrochemical solar cells do present distinct processing requirements. Unlike solid-state cells, there are no severe specifications to be observed for crystal quality or doping procedures. Clean room conditions, and high-temperature treatments such as in diffusion furnaces are unnecessary. The cells are fabricated by simple tape casting of a slurry of titania powder on a conducting glass surface, then sintering to form a coherent transparent film. This heat treatment is limited to only 450 °C in order to avoid high-temperature phase changes. Thereafter the chemisorption of the dye is by immersion in a solution. The preferred electrolyte is an iodine/iodide (I^-/I_3^-) redox couple in an organic solvent. The counter-electrode is a second conducting glass surface, or even a polymer-supported carbon film, with a small amount of high-dispersion platinum as catalyst.

Of course, an adequate interconnection technology to array individual photoelectrochemical cells in modules is an imperative for successful large-scale solar energy conversion. The dye-sensitised device shares with amorphous

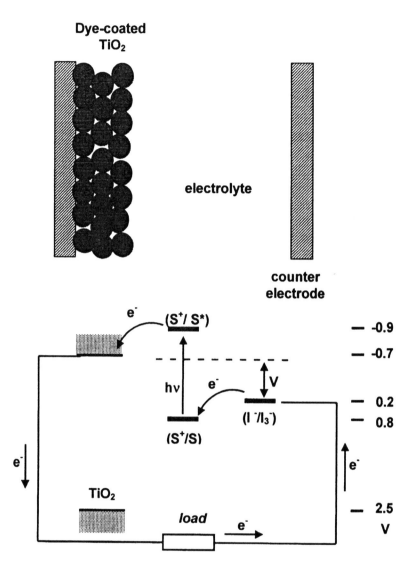

Fig. 7.17 Charge transfer mechanisms in the dye-sensitised nanocrystalline photoelectro-chemical cell. The sensitiser (S) is excited by the energy of the absorbed photon, then relaxes by electron injection into the semiconductor layer. The charged dye molecule is neutralised by the redox system, itself regenerated at the counter-electrode by electrons passed through the load. Potentials are referred to the standard calomel electrode (SCE)

silicon and other thin-film devices supported on transparent conducting oxides, the geometrical restriction that every point on the photoactive surface must be less than 1 cm from a current collector or other interconnection device. This is a consequence of the limited conductivity of the conducting glass. The modules for large-area solar energy conversion therefore present a common format with other thin film technologies, the individual cells forming strips interconnected

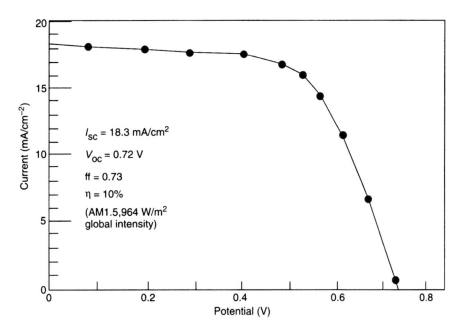

Fig. 7.18 Current–voltage characteristic of dye-sensitised electrochemical photovoltaic device, measured under AM 1.5 spectrum with global irradiance 964 W/m²

along their lengthwise edges. There is the additional requirement in these devices that the interconnect structures also form seals for the fluid electrolyte, and with quality control requirements all the more rigorous because of the electrical potential photogenerated between adjacent cells. However, several sealing concepts under development can claim significant progress, for example, by the use of low-temperature glass frits or polymer materials. However, one particular route to confront the long-term sealing reliability problem is to replace the fluid electrolyte with a gel or solid. In that case, if charge transport is to be maintained by ionic mobility, there is a difficult compromise necessary between that mobility and the electrolyte viscosity. A further complication is that the desirable low viscosity is often associated with low boiling point and therefore a significant vapour pressure within the cell under the temperature conditions of full-sun energy conversion. It is in this connection that the all solid-state sensitised heterojunction solar cell concept offers a promising alternative development strategy.

7.3.2.4 Dye-sensitisation in heterojunctions

In the proposed solid-state variant of the dye-sensitisation concept, it should be remembered that while the dye is electrochemically active, it is not itself electrically conducting. As in the electrochemical photovoltaic device, therefore, it must be present only as a monomolecular film between two conducting

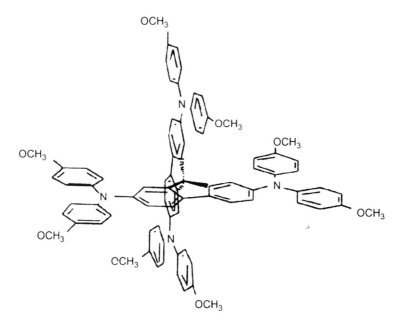

Fig. 7.19 Molecular structure of a recently-developed hole-conducting (p-type) organic compound, a spirobifluorine, for use in sensitised heterojunction cells

phases. The n-type component of a heterojunction device can, of course, be the nanocrystalline titania layer already described, with the dye chemisorbed on it. Evidently, in the photoelectrochemical version of the sensitised cell the liquid electrolyte penetrates into the porosity, thereby permitting the intimate contact with the dye monolayer necessary for charge exchange with the redox system. It is not so immediately evident that an interpenetrating network of two conducting solids can be established for an immobilised molecule at their interface to exchange charge carriers with both of them. However, prototype devices with organic charge transfer materials as the second contacting phase are promising (Bach *et al.*, 1998).

These can be deposited by spin coating a solution on to the titania surface. One such charge transfer material currently used is a spirobifluorene, proprietary to the Hoechst company (Lupo and Salbeck, 1996) whose structure is shown in Fig. 7.19. If this material functions in the cell as a p-type conductor with holes as the mobile charge carriers, the device can be called sensitised heterojunction. If, however, spirobifluorene molecules accept positive charge to become cations, the charge transfer mechanism within this solid organic phase can still be considered a redox equilibration, and the device is photoelectrochemical. While the functionality of the system for photovoltaic applications is unaffected, this distinction may have patent implications. It will also be noted that the dye now exchanges majority carriers with each phase, so the previous comments about suppression of recombination still apply.

7.3.3 Photocatalysis and photochemistry

A final application of semiconductors for the exploitation of solar energy in photochemistry which will be mentioned briefly are suspended particulate semiconductors. Each grain can be regarded as a photoelectrochemical cell on which surface irregularities or deliberately added catalysts can be preferential for electron or hole exchange with the supporting medium. When photo-excited, semiconductor suspensions can initiate or sustain chemical reactions. Photography is essentially a localised photoelectrochemical reaction on silver halide semiconductor particles immobilised in a gelatin film. Other practical applications of these particulate semiconductor photosystems are being investigated, particularly in environmental engineering where they can promote the oxidative elimination of pollutants, particularly toxic products incompatible with conventional biological waste-water treatment. In certain cases the photo-dissociation products can themselves be of economic interest, for example the hydrogen and elemental sulphur produced by photoelectrolysis of sulphide liquors in contact with illuminated cadmium sulphide particles. At least the outflow of a photochemical reactor can be safely discharged, as in the photo-oxidation of cyanide on zinc oxide particles, which ultimately yields ammonia and dissolved carbonate, via the less toxic intermediate cyanate. A wide range of organic residues, including chloro-fluoro-carbon (CFC) compounds and carcinogenic aromatics can be mineralised or oxidised to innocuous products by photochemical reaction in contact with titania suspensions. There are even cases of environmental photochemistry in contact with gas, rather than liquid, as for example the photo-oxidation of vapours of paint solvents in the automotive industry. In future solar engineering, therefore, it is likely that solar photochemistry may provide a third component alongside photovoltaics and solar thermal energy conversion.

Box 7.1 Electrolytes and electrochemical cells

An electrolyte is a medium in which charge transport is not an electronic process through the mobility of free electrons or holes, but in which the charge is associated with mobile atoms or molecules as ions. Although normally considered to be liquids, solid electrolytes are also known. Examples include the ceramic, zirconia, where the charge carrier for conduction is the oxygen ion at elevated temperatures, or the fluorinated sulphonate polymers in which hydrogen ions are mobile. These materials are used in fuel cells and in industrial electrochemistry, and are discussed more fully in Section 7.4. Charge is transferred to and from the conducting electrolyte at the interfaces with electrodes. The cathode is the electrode from which an electron is transferred to charge an ion negatively in a process recognised chemically as reduction, whereas the anode is the site of the oxidation reaction by accepting electrons from ions to leave them more positively charged. Positive ions such as oxidised metals, reducible at a cathode, are called cations. Anions are negatively charged species, oxidisable at an anode. All conduction through electrolytes therefore requires two complementary electrochemical reactions to take place – oxidation and reduction – at the respective electrodes. The system comprising the two electrodes and the electrolyte constitutes an electrochemical cell. The electrodes are therefore identified by their electrochemical functionality rather than, as is often thought, by the polarity of the potential difference between them.

Within an electrochemical cell it is possible to measure only this potential difference between the electrodes. It is therefore arbitrary but convenient to select one electrochemical reaction as the standard potential in electrochemistry. The reaction chosen is the evolution of hydrogen from aqueous solution, one half of the familiar process of electrolysis of water. The conditions are also standardised: the concentration of hydrogen ions in solution in the electrolyte is 1 mole per litre and the pressure of hydrogen gas equal to 1 atmosphere at $25\,°C$. In aqueous solution such a concentration can be obtained only in a strongly acid solution when pH $= 0$. Neutral water, by contrast, has a pH of 7. The reference potential is reached when an effective nonpolarised electrode such as platinum is in equilibrium with gaseous hydrogen and the pH 0 electrolyte so that no current flows. This system is referred to as the normal hydrogen electrode (NHE). Experimentally, it has been determined that the NHE potential lies some $4.5\,V$ negative of the vacuum level determined for solids, thereby providing a basis for relating the electrochemical and solid-state energy scales (see Memming, 1984). Given the typical work functions of semiconductor materials, it follows that many will find their Fermi level at an energy typical of aqueous-phase electrochemical processes. In electrochemistry, it is also convenient to use a single stable device which does not require hydrogen gas. Such a convenient secondary standard is the calomel electrode, based on a redox process in mercury chloride.

The electrochemical processes at the electrodes may involve the same or different ionic species. In the first case, a cation is reduced at one electrode and then diffuses to the other where it is re-oxidised. Such a cell is regenerative: the concentration of reagents in the cell remains invariant and the only effect is the overall charge conduction between the electrodes. The electrolyte in a regenerative cell represents a redox (**RED**uction/**OX**idation) system. However, if unrelated reactions take place (electrolysis or electrosynthesis) the consequence is an exchange between electricity and chemical energy. The most familiar example is clearly the separation of water into its component elements. In this case, the hydrogen evolution at the cathode by reduction of positively charged hydrogen ions (H^+) is complemented by the oxidation of water or hydroxyl anions (OH^-), and the appearance of gaseous oxygen at the anode. The quantity of electricity required to produce 2 grams (1 mole) of hydrogen is simply

$$2qN = 2F = 1.93 \times 10^5\,C$$

since for each water molecule H_2O, it takes two electrons to release one hydrogen molecule H_2. The minimum potential difference required is

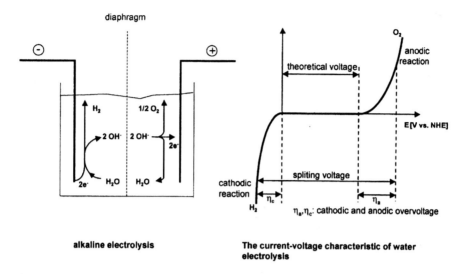

Fig. B7.1 The principles of water electrolysis. The voltage is referred to the normal hydrogen electrode, the zero point of the electrochemical scale

determined by the equilibrium potential of the reactions at each electrode and is, in the present case, equal to 1.23 V (Fig. B7.1).

The table of standard potentials for electrochemical processes of atoms or molecules constitutes the electrochemical series, and is the basis for selection of materials for batteries and other electrochemical energy conversion and storage systems (Handbook of Physics and Chemistry, p. D-162). In practical systems an overpotential must be applied to accommodate eventual loss mechanisms such as the formation of intermediate products and kinetic limitations, and to pass a significant current through the electrochemical cell. Hence practical electrolysers require a potential difference of 1.6–2.0 V per cell, while oxygen – hydrogen fuel cells generate about 1.0 V, somewhat lower than the difference of equilibrium potentials.

	cathode $(-)$	anode $(+)$	Remarks
Acid (pH 0)	$2H^+ + 2e^- \Rightarrow H_2$	$2H_2O \Rightarrow 4H^+ + O_2 + 4e^-$	Fig. B7.1
Alkali (pH 14)	$2H_2O + 2e^- \Rightarrow H_2 + 2OH^-$	$4OH^- \Rightarrow 2H_2O + O_2 + 4e^-$	Fig. B7.1

Reference

Handbook of Physics and Chemistry, CRC Press, Boca Raton, FL, 63rd edition.

Summary

In conventional solid-state photovoltaics, a semiconductor provides the medium for optical absorption and a junction to or within the semiconductor is the site for charge separation. The mobile charge carriers are negatively-charged electrons or their complement, electron deficiencies interpretable as positively-charged holes. However, the absorbing component of the photosystem is a dye, chosen so that the photogenerated higher energy state of the molecule can lose that energy by electron transfer to a substrate. The original neutral state is subsequently recovered by redox electrochemical reaction with a contacting electrolyte medium in which the charge carriers are ions. This alternative photovoltaic concept is now a scientifically established system, even though technically and commercially such devices are still in the development and prototype stage. The significance of the establishment of the nanocrystal-line electrochemical photovoltaic device is not just the challenge it represents to the conventional solid-state technology, but the fact that the monopoly of the all-semiconductor solar cell is broken, and the way opened for the acceptance and development of novel ideas and processes in photovoltaics.

7.4 ALTERNATIVE STORAGE: THE HYDROGEN ECONOMY

7.4.1 Introduction

As we have amply discussed throughout this book, the energy generated by a photovoltaic generator usually needs storage to compensate for the intermittent nature of solar radiation. The conventional energy storage by batteries was discussed in Chapter 4. In this section we describe how solar energy can be stored in an alternative chemical form, by hydrogen production from water. Indeed, as we shall see presently, this technology represents more than energy storage: it offers a new form of energy economy which complies with the current environmental requirements.

The advantages of using hydrogen as energy storage are summarised in Table 7.5. An important point to note is that the use of hydrogen does not generate harmful emissions to the environment. In particular, it does not produce carbon dioxide which contributes to the greenhouse effect. Pure carbon burns to CO_2, creating 0.34 kg of CO_2 per kWh when coal is used (kWh here refers to the thermal energy). Hydrocarbons burn to form CO_2 and water. The CO_2 emission is here about 0.27 kg in the case of oil and 0.2 kg for natural gas. In comparison, hydrogen combustion is CO_2 free.

In the course of the industrial development, there was a change from coal to other fuels with lower CO_2 emissions. It is envisaged that this trend will continue through the intensified use of natural gas with a high hydrogen content to the exclusive use of hydrogen which burns without CO_2 emissions. The H/C ratio of coal is about 1:1 whilst for the natural gas, it is 4:1. When hydrogen is used on a large scale, the H/C ratio of the energy carriers will be greater than 4:1.

It is foreseen that the hydrogen economy would consist of the following components (Fig. 7.20).

Table 7.5 Advantages of hydrogen as energy carrier

- *Established technology*. Hydrogen generation by electrolysis is a well-known process. Hydrogen is one of the oldest industrial gases and its handling is well understood.
- *Compatible with photovoltaic generation*. Water electrolysis requires a low-voltage DC source.
- *Ease of use and storage*. Hydrogen can be used directly for electricity, heat and power generation, and can be stored with minimal losses.
- *No global environmental risks*. Hydrogen is environmentally neutral. The reaction product is water (no emissions of CO_2, CO, SO_2, hydrocarbons, dust or ash). In addition to water, only traces of NO_x are formed in the combustion process.
- *Feasible generation and transport*. Hydrogen can be generated in solar farms in the sun belts of the earth, and transported to the consumer by pipelines or tankers.
- *Existing infrastructure can be used*. Natural gas can be successively replaced by hydrogen which will be fed into the existing natural gas pipeline system (in former times, town gas was used with a content of 50–60% hydrogen).

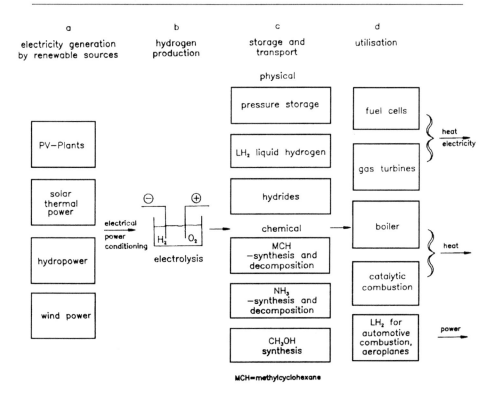

Fig. 7.20 Components of the hydrogen economy

- *Electricity generation* by photovoltaic plants or other renewable sources. It is important to choose an energy supply of the future that will not damage the environment. This can only be based on renewable energy sources.
- *Production of hydrogen* by electrolysis (section 7.4.2).
- *Transport and storage of hydrogen.* The physical and chemical methods for local storage systems, automotive storage and intercontinental transport are discussed in section 7.4.3.
- *Utilisation of hydrogen,* for example, heating, electricity generation in fuel cells and gas turbines, or as transportation fuel (section 7.4.4).

7.4.2 Generation of hydrogen

7.4.2.1 Electrolysis

Hydrogen can easily be generated by electrochemical splitting of water (electrolysis – see Box 7.1 Electrolytes and electrochemical cells). There is a number of ways in which electrolysis may be implemented in practice (Fig. 7.21). In the most advanced version of the conventional or *alkaline water electrolysis,* the electrolyte is 30% KOH where the ion transport takes place via OH^- and K^+.

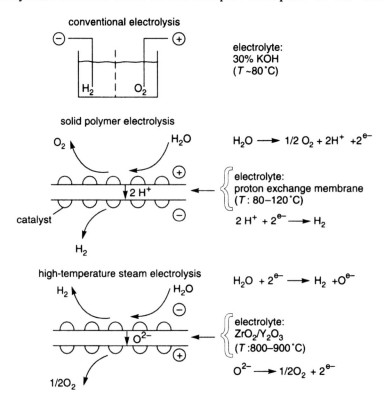

Fig. 7.21 The different types of electrolysis

A diaphragm is inserted between the two electrodes and prevents mixing of the hydrogen and oxygen gases. The electrodes are usually made from nickel, and the electrolysis proceeds at a temperature of 80 °C.

In the electrolysis based on *solid polymer electrolysis*, the electrolyte is a proton-exchange membrane which conducts H^+ ions. The operating temperature is usually between 80 and 120 °C.

High-temperature electrolysis (at temperatures 800–900 °C) is not available yet, but promises the greatest energy efficiency, converting more than 90% of the electric energy. The solid electrolyte, based on ZrO_2/Y_2O_3, conducts here the O^{2-} ions.

There are several requirements on the interconnection between the photovoltaic generator and the electrolyser, as well as on the electrolyser itself. The principal ones are:

- good dynamic response at intermittent load,
- high gas purity (H_2/O_2 mixtures form explosive gases),
- optimisation of power conditioning between the PV generator and the electrolyser.

The energy efficiency of a water electrolyser can be determined from the following arguments. An electrolytic cell normally operates at current densities of a few hundred mA/cm^2. This is a compromise between economy and energy efficiency. The energy consumption for a conventional electrolyser is about 4.8 kWh/Nm^3 (kWh/norm cubic meter of hydrogen at normal conditions of 273 °K and 0.1 MPa) but for an advanced alkaline technique this value is down to 3.9 kWh/Nm^3. The theoretical value with 100% efficiency is 3.0 kWh/Nm^3 (referred to as the lower heating value of hydrogen) giving energy efficiency of an advanced electrolyser as $(3.0/3.9) \times 100 = 77\%$.

The technology of continuous water electrolysis is well developed. However, discontinuous electrolysis powered by intermittent sources (photovoltaics or solar thermal generators) requires more research and development. During solar operation, the electrodes have to cope with a widely fluctuating load. This behaviour affects the activity of the electrocatalysts. Protection potentials have to be applied to the electrodes when there is no irradiation (at night). Further design criteria for solar operation of an electrolyser are low specific energy consumption, high gas purity, and good current efficiency at partial load, as well as a good dynamic response at intermittent load in the entire current density range.

Since the photovoltaic generators and electrolysers are modular, a variety of electric connections are possible. The simplest connection between the PV generator and electrolyser is by direct wiring with different string connections. Higher efficiencies are attainable by using a power-conditioning DC/DC converter with full MPP tracking. Up to 95% of the generator power can then be directed to the electrolyser.

7.4.2.2 Other hydrogen generation technologies

An alternative concept for hydrogen production is a direct conversion of light energy into chemical energy in a photo-electrochemical reaction (analogous to photosynthesis in biology). In contrast with the two-step photovoltaic electrolysis system, the photo-electrochemical reaction is a one-step process. An illuminated, hydrogen-evolving semiconductor photo-cathode can be combined with a photoanode to form a photo-electrolytic cell. In principle, this concept represents a simpler technological method, but, due to corrosion processes at the semiconductor electrodes, photo-electrochemical water-splitting systems are not available at present. Other direct solar-to-chemical energy conversion methods such as biological and thermal processes are still in the research stages and will not be available in the near future.

7.4.3 Transport and storage of hydrogen

The hydrogen storage technology can be divided into physical and chemical methods.

7.4.3.1 Physical methods

The simplest storage system for hydrogen is to fill pressure vessels and gas tanks with *gaseous hydrogen* (GH_2). This system plays an important role for the stationary storage of natural gas and hydrogen, and in pipeline systems. Because of the low energy density and high weight of the pressure vessel as well as the necessary safety precautions, this system is not suitable for automotive storage in personal passenger cars. For intercontinental hydrogen transport from the sun belts to the consumer, pipelines can be used in the same way as with the oil and natural gas transport today. This large transport system is also a storage medium which is able to buffer between the intermittent hydrogen generation and energy consumption.

The energy density of hydrogen can be dramatically increased by liquefaction. The temperature of *liquid hydrogen* (LH_2) is about $-253°C$ which requires the use of well-insulated tanks. In comparison with the density of GH_2 at 1 bar ($90 \, g/m^3$), the density of LH_2 is $71 \, kg/m^3$. Liquefaction of hydrogen, however, consumes large amounts of energy. One third of the inherent energy of the gaseous hydrogen is used for liquefaction. LH_2 therefore retains only about 55% of the electric energy used to produce it. In the hydrogen economy, LH_2 will be used for intercontinental shipping in tankers, for hydrogen-powered airplanes, and for automotive storage.

Certain metals, in the form of *hydrides*, can store hydrogen embedded in their molecular lattice. These metal hydrides are of interest for hydrogen-powered cars (automotive combustion or electricity generation in fuel cells). The advantage of this system is a high charge and discharge efficiency, but the disadvantage is the high weight of the metal hydride storage.

7.4.3.2 Chemical storage methods

In addition to liquefaction, chemical storage is of interest for the intercontinental tanker transport of hydrogen. In many applications, the handling of liquid fuels at ambient temperature is easier than the low-temperature storage (see Fig. 7.20).

When combined with hydrogen, chemicals like toluene form new compounds. The reaction product is methylcyclohexane (MCH) in the case of toluene:

$$C_7H_8(\text{toluene}) + 3H_2 \rightarrow C_7H_{14}(\text{MCH}) \tag{7.10}$$

MCH is liquid at ambient pressure and temperature and can be stored and transported in existing oil tanks and ships. After transport, the hydrogen carrier has to be dehydrogenated and the hydrogen is ready for use. Toluene has to be shipped back again for hydrogenation in a two-way transport. Another possible chemical storage of hydrogen is by ammonia production:

$$N_2 + 3H_2 \rightarrow 2NH_3 \tag{7.11}$$

Liquid ammonia is contained unpressurised at $-33°C$, or at the normal temperature at a pressure of 10 bar. The low-temperature transport here does not represent the same problem as for liquid hydrogen. The overall efficiencies for hydrogen storage in the form of MCH (59%) and NH_3 (61%) are slightly higher than the LH_2 storage (55%).

The use of solar hydrogen as a chemical raw material for the production of synthetic carbonaceous fuels (synfuels with a high H/C ratio, e.g. methanol) is one of the most important topics in solar hydrogen storage research. One molecule of methanol contains four hydrogen atoms. Compared to other hydrogen energy carriers like methylcyclohexane or ammonia, methanol is easy to transport and handle, and is ready for use without any chemical decomposition processes. The methanol technology as an energy carrier is fully developed, and storage of liquid methanol is possible at the ambient temperature. Methanol offers a 1.8 times higher volume energy density than liquid hydrogen. An additional economic advantage would be the possibility of using the existing infrastructure of gas stations and transport systems for automotive combustion.

The combination of fossil and solar sources is feasible, for example, by combining synthesis gas with solar hydrogen to produce methanol:

$$(H_2 + CO)_{\text{fossil}} + H_2 \rightarrow CH_3OH \tag{7.12}$$

The synthesis of methanol can be achieved by a catalytic chemical process between gaseous CO_2 and H_2:

$$CO_2 + 3H_2 \rightarrow CH_3OH + H_2O \tag{7.13}$$

The recycling of CO_2 will help to prevent the rise in the atmospheric concentration of this gas. In the race to develop a cleaner-burning fuel from renewable sources, methanol is a practical option for solar hydrogen storage. Calculated values of the overall energy efficiency of the synthesis, however, are slightly lower than in LH_2 generation.

7.4.4 Hydrogen utilisation technologies

Hydrogen has a wide range of uses for heat, power and electric generation (see Fig. 7.20). Some examples are listed below.

Fuel cells. In the hydrogen economy, fuel cells will play a major role. The energy efficiency of electricity generation directly from chemical energy is not limited by the Carnot process, and values in excess of 60% can be realised. The principle of a fuel cell is the reversal of electrolysis. Hydrogen is oxidised at the anode (which is here the negative electrode) and oxygen is reduced at the cathode (positive electrode). Figure 7.22 shows the current – voltage behaviour of a fuel cell which should be compared with that of an electrolytic cell (Fig. B7.1):

$$\text{anode:} \qquad H_2 \rightarrow 2H^+ 2e-$$
$$\text{cathode:} \qquad \tfrac{1}{2}O_2 + 2H^+ + 2e^- \rightarrow H_2O$$
$$\text{overall reaction:} \quad H_2 + \tfrac{1}{2}O_2 \rightarrow H_2O$$

The reaction product is water without any pollutants.

Catalytic burners. In comparison with flame combustion, catalytic combustion of hydrogen exhibits an inherent advantage of high reaction efficiencies combined with very low NO_x emissions. During flame-free combustion, hydrogen reacts with air when in contact with solid catalysts at relatively low temperatures. Catalytic heaters operate in a range from the ambient temperature to several hundred degrees Celsius.

Hydrogen for steam generation. Steam generators can be used for peak load in conventional electric power plants and for industrial processes. The power can be increased from 0 to 100 MW in 1 s by a hydrogen/oxygen steam generator based on rocket technology.

Hydrogen in aviation. The advantage of using hydrogen in aviation is the weight of LH_2. With the same energy content, LH_2 weighs only one third compared to conventional aircraft fuels. The take-off weight is reduced by about 30% for the same payload and range.

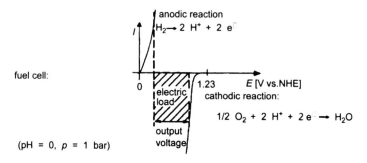

Fig. 7.22 Operation of the fuel cell

Hydrogen as raw material. Hydrogen is also an important raw material in industrial synthesis processes (for example, ammonia synthesis for fertilisers, hydrogenation in petrochemical industry, methanol synthesis, reduction processes in metallurgy, etc.)

Summary

Hydrogen economy combined with photovoltaic generation offer the possibility of energy technology without carbon dioxide or other environmentally damaging emissions. The hydrogen generation by electrolysis is a well-known process. The storage and transport are well understood and can build on the existing infrastructure of pipelines and tankers. These aspects, together with hydrogen utilisation technologies were discussed in detail, highlighting future potential of the technology.

BIBLIOGRAPHY AND REFERENCES

BACH, U., LUPO, D., COMTE, P., MOSER, J. E., WEISSÖRTEL, F., SALBECK, J., SPREITZERT, H. and GRÄTZEL, M., *Nature*, **395**, 1998; 583.
BECQUEREL, A. E., *C.R. Acad. Sci.*, **9**, 1839; 561.
COOKE, D., GLECKMAN, P., KREBS, H., O'GALLAGHER, J., SAGIE, D. and WINSTON, R., Sunlight brighter than the sun, *Nature* **346**, 1990: 802.
DESILVESTRO, J., GRÄTZEL, M., KAVAN, L., MOSER, J., and AUGUSTYNSKI, J., *J. Am. Chem. Soc.* **107**, 1985; 2988.
FUJISHIMA, A. and HONDA, K., *Bull. Chem. Soc. Japan* **44**, 1971; 1148.
GERISCHER, H. and TRIBUTSCH, H., *Ber. Bunsenges. Phys. Chem.* **72**, 1968; 437.
LUPO, D. and SALBECK, J., *Intern. patent PCT/EP96/03944.*
LUQUE, A., *Solar Cells and Optics for Photovoltaic Concentration*, Adam Hilger, Bristol, 1989, p. 111.
MEMMING, R. *Progress in Surface Science*, **17**, 1984: 7.
MEMMING, R. *Topics in Current Chemistry*, **143**, 1988: 81.
MOSER, J., *Monatsh. Chem.*, **8**, 1887: 373.
WELFORD, W. T. and WINSTON, R. *High Collection Nonimaging Optics*, Academic, New York, 1989, p. 50.
WEST, W., *Proc. Vogel Centennial Symp., Photogr. Sci. Eng.*, **18**, 1974: 35.
WINSTON, R. Non-imaging optics, *Scientific American* **264**(3) 1991: 52–57.

SELF-ASSESSMENT QUESTIONS

Large PV projects

True or false?

1. Large PV projects are measured on the scale of:
   (a) hundreds of megawatts
   (b) hundreds of kilowatts.

2. Large PV projects are generally more cost effective in:
   (a) grid-connected applications
   (b) stand-alone configurations.

3. Under favourable climatic conditions the unit cost of electricity generated by a large-scale renewable energy source is only slightly higher than that from a conventional power plant. This statement generally applies to:
   (a) wind power
   (b) PV solar power.

4. Under favourable circumstances PV systems may be more cost-effective and compete with conventional power sources. This statement generally applies to:
   (a) small-scale stand-alone applications
   (b) large-scale grid-connected PV power plants.

5. For PV applications in the Third World the following overall energy system conception should be preferred:
   (a) concentrated large-scale PV power plant feeding users through a conventional distribution grid
   (b) many diffused small-scale PV systems feeding relevant users locally.

6. According to the principle of 'economy of scale' a large production facility presents lower unit production costs (cost/unit) than a small production facility of otherwise same characteristics. Generally this principle applies also to:
   (a) conventional large-scale (bulk) power plants
   (b) diesel power plants
   (c) PV module manufacturing
   (d) PV systems (applications).

7. In grid-connected PV power plants the installation of a battery yields significant advantages since it allows accumulation of excess energy during the day and supply to users (of the grid) also during the night.

8. In large-scale PV systems a high DC bus voltage generally yields higher system efficiency.

9. Considering the high-reliability standards reached by modern computers, the best strategy for the automation of a PV plant is to provide for this purpose:
   (a) a centralised and powerful process control system
   (b) a distributed control system consisting of independent functional blocks and control loops.

Photovoltaics under concentrated sunlight

1. A choice is to be made between two photovoltaic systems:
   (i) conventional panels costing \$500 per m^2, and
   (ii) a concentration system using lenses costing \$200 per m^2 of area from which sunlight is collected, together with cells, whose area is only 1/100th of that area, costing \$2000 per m^2 of cell area.

Each system produces the same average power output for a given collection area. On the basis of these figures, which is the cheaper? What other consideration might affect the choice in practice?

2. A perfectly designed lens could theoretically concentrate sunlight to an infinite extent—is this true or false?

3. What merits and drawbacks do line-focus concentrators have as compared to point-focus ones?

4. What do you understand by an *ideal concentrator?*

5. Why is a sun tracker more important for concentration systems than for ordinary photovoltaic panels?

Hydrogen economy

1. What are the main advantages of hydrogen as a solar energy carrier?

2. What are CO_2 emissions of different fossil energy carriers?

3. What are the major components of a solar hydrogen economy?

4. What are the reactions at the electrodes during water splitting in a conventional electrolysis cell?

5. What is the theoretical and practical splitting voltage?

6. How does one calculate the energy efficiency of a water electrolyser?

7. What are the main requirements on the electrolyser, and on the interconnection between a PV generator and an electrolyser?

Answers

Large PV projects

1a, False; 1b, True; 2a, True; 2b, False; 3a, True; 3b, False; 4a, True; 4b, False; 5a, False; 5b, True; 6a, True; 6b, True, 6c, True, 6d, False; 7, False; 8, True; 9a, False; 9b, True.

Photovoltaics under concentrated sunlight

1. The problem is to work out the cost of the second system per m^2 of collection area and compare it with the conventional system. For every m^2 of collection area we have in (ii) 0.01 m^2 of cells, costing $20, which added to the cost of the lenses give 220/m^2 in total. This is cheaper than the conventional system. However, the sun-tracker would add to the cost of the second system and might affect the reliability.

2. False: the concentration is limited by thermodynamic considerations, however good the lens is.

3. Line-focus concentrators are generally easier to construct since they can more often make use of flat sheets of material bent into shape. Also, they may be used with tracking about one instead of two axes. However, the concentration ratios they can produce are much lower.
4. Ideal concentrators produce concentration ratios in relation to the angle within which they collect light, as high as the theoretical maxima of equation (7.7) (or equation (7.8) for concentrators of linear symmetry). No energy is lost in them, i.e. all light entering within the collection angle is transmitted to the solar cell.
5. Sun-trackers are used to keep solar concentrators pointed towards the sun as it appears to move through the sky. Unlike ordinary photovoltaic panels, concentration systems only accept light within a certain range of angles and so must be orientated correctly with respect to the sun.

Hydrogen economy

1. See Table 7.1.
2. Burning of coal, $0.34\,kg\ CO_2$ per kWh_{th}, 0.27 for oil, and 0.2 for natural gas.
3. Electricity generation by a renewable source, hydrogen production by electrolysis, transport and storage of hydrogen, followed by utilisation.
4. $2H^+ + 2e^- \rightarrow H_2$ at the cathode

$$H_2O \rightarrow \frac{1}{2}O_2 + 2H^+ + 2e^- \text{ at the anode}$$

The overall reaction is $H_2O \rightarrow H_2 + \frac{1}{2}O_2$
5. Theoretical splitting voltage is $1.23\,V$. In practice, voltage greater than $1.6\,V$ is required.
6. Energy efficiency

$$= \frac{\text{Lower heating value of hydrogen}(= 3.0\,kWh)}{\text{Energy consumption in kWh per norm } m^3 \text{ of hydrogen}}$$

7. Good dynamic response at intermittent load, high gas purity, and MPP tracker for best operation.

Index